THE HIGH FRONTIER

About the Author

Gerard K. O'Neill is Professor Emeritus of Physics at Princeton University, and President of the Space Studies Institute. He founded the Geostar Satellite Corporation, and is Chairman of O'Neill Communications, Inc. In 1985 he was appointed by President Reagan to the National Commission on Space. He is the author of *2081: A Hopeful View of the Human Future*, and *The Technology Edge: Opportunities for America in World Competition*.

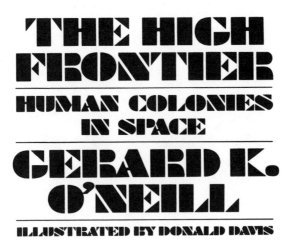

THE HIGH FRONTIER

HUMAN COLONIES IN SPACE

GERARD K. O'NEILL

ILLUSTRATED BY DONALD DAVIS

Space Studies Institute Press
Princeton, New Jersey
1989

Library of Congress Cataloging in Publication Data
O'Neill, Gerard K.
The high frontier.
Includes bibliographical references and index.
1. Space colonies. I. Title.
[TL795.7.O53 1982]   629.44'2   82-45263
ISBN 0-9622379-0-6

To
Edward
and
Dorothy

CONTENTS

PREFACE

I first read *The High Frontier* many years ago, long before I ever imagined that I would someday orbit the Earth myself. The book's central premise—that much of the general knowledge and technology needed to design a human habitat in space were already in hand—made quite an impression upon me in 1976.

I recall, too, that the book provoked quite a flood of questions for me: Could such a future really be so close at hand? Would life in these colonies really be different than life in human societies on Earth? Why wasn't progress in this direction more evident, if so many pieces of the puzzle lay so near at hand? Gerry addresses these issues, noting that politics, finance and national will usually set the pace for our space plans to a greater degree than do technical issues. At the same time, however, he staunchly refuses to give in to the cynicism or pessimism to which they all too often give rise. This allows him to focus instead on the solvable, technical challenges involved and, thus, to give us a vision of an exciting future towards which we can work. I think he does a great service by presenting this vision in great detail and disseminating it broadly, provoking questions such as mine in the minds of present and future engineers, government leaders, financiers and space citizens. It is essential to maintain a positive vision of the future, from which to draw our goals, the motivation to pursue them, and the compulsion to meet the complex human challenges we will face along the way.

As you read this book, I hope you will come to share some of this vision and will consider some of the problems that confront us now, as we strive to become a spacefaring people in the next century. I trust you will enjoy, as I did, the richly-detailed descriptions Gerry gives of what some future homes in space might be like. Of course, whether or not he has gotten the layout of the homes and gardens, or solar panels and mirrors, exactly right is not the vital point. The key questions are: What will the motives and incentives actually be that someday drive significant human migration into space? What key technologies, or financial and political arrangements, could be developed today to enable this? And a question I hope many readers, especially young ones, will ask themselves: What role can I play in mankind's progress towards the High Frontier?

I, for one, believe that increasing both human activity in space and automated exploration of our solar system will teach us many lessons, and will compel us to develop many technologies, that will be important, in some cases perhaps crucial, to the well-being of human societies on Earth, as well as to our possible descendants in space. Can there be any better motivation for persevering through the challenges ahead of us on the space frontier? Well, perhaps just one: the view of Earth out the window of my future living room. . .

Kathy Sullivan
Astronaut
1989

INTRODUCTION

Following the publication of *The High Frontier*, the existing thrust of our human species outward to the new world of space continued through the work of many nations. Western Europe, Japan, India and the People's Republic of China all developed independent capabilities for launching satellites and spaceprobes. Western Europe and China in particular also began working toward manned spaceflight, and Japan is now planning to launch spacefarers on its "Hope" space plane, a reusable glider.

The space activities of the U.S.S.R. and the United States moved along diverging lines, reviewed in the new chapter which completes this book. The High Frontier concept, using the material and energy resources of space to improve the human condition on Earth and to build colonies in space, was developed and supported by the work of many people. The basic arguments and conclusions of the High Frontier concept were buttressed by almost a dozen studies and reviews, done mainly by NASA and its contractors. In 1977 a new, independent, citizen-supported organization, the Space Studies Institute (SSI) was formed. The Institute assumed the task of funding the basic research necessary to our attainment of the High Frontier. Its research included successful projects on mass accelerators and the processing of lunar materials. Institute research into economical ways of manufacturing large products in high orbit, such as solar power satellites, showed that manufacturing in space from lunar materials can be highly beneficial to life on Earth, and profitable as well.

Space colonies, our main topic, are now seen as an inevitable result of the large-scale development of space resources. They are essential, because they turn space into much more than a location for transient occupancy. In the modern view space will become a rich, new, Earthlike environmental range for humanity, bathed in continuous free energy.

Detailed engineering studies verified that the "Island One" spherical geometry for space colonies ranked highest in simplicity, ruggedness, economy and safety among Earthlike colony designs. The L5 orbit for colonies was shown to have no strong advantage over other high circular orbits. "L5," as used here, should therefore be taken as shorthand for any high circular orbit about the Earth or the Sun.

When the Space Studies Institute decided to publish a new edition of the High Frontier, naturally I was tempted to edit the main text, in order to reflect progress since the book was written. As a matter of principle I resisted that temptation. While this introduction is new, a few introductory and closing lines of Chapter 1 are edited, and Appendix 2 has been replaced by the new "Perspective: a View from 1988," all of the book from Chapter 2 through Appendix 1 remains exactly as it was written originally. My reasoning is that this work both recommends our future course and predicts what the result will look like. To judge its validity, you as its reader should be free therefore to review the original as the years go on. My own re-reading of the book, subjective to be sure, reassures me that its logic will endure, and that future history will develop in broad outline as it predicts. I leave to readers still farther away in time what I hope will be the pleasure of comparing future reality to my perception of it seen from the world of 1976.

In the last eventful years one of my great joys has been meeting and coming to know well a number of highly talented individuals, who work together toward our reaching the High Frontier. I cannot name several without omitting still others who deserve thanks. But leading all who work toward the High Frontier, I thank especially Gregg Maryniak, the Senior Associates, staff, members and volunteers of the Space Studies Institute, and all who work with them, for their creativity, their dedication, and their continuing hard work.

Gerard K. O'Neill
Princeton, New Jersey
1988

THE HIGH FRONTIER

1

A LETTER FROM SPACE

In 1974 a new concept to improve the human prospect entered the arena of open discussion. Its thrust is to open for our use new sources of energy and materials while preserving our environment. First it was known as "space colonization," but now, as it is discussed with increasing seriousness in the circles of government, business, the universities, and

1

the press, we tend to use for it less dramatic names: "space manufacturing," or "high-orbital manufacturing."

The concept of human habitation in space is, of course, a very old one; in some form it can be traced back to the early days of science, and even earlier to mysticism. It has been a theme for fiction over several decades, and at least one fictional discussion of an inhabited artificial satellite, by Edward Everett Hale, was written during the latter half the nineteenth century. The Russian schoolmaster and physicist Konstantin Tsiolkowsky foresaw certain elements of the space community concept with remarkable clarity. In a novel, *Beyond the Planet Earth*, written about 1900 and published some twenty years later, Tsiolkowsky set his space travelers to work, on their very first voyage, constructing greenhouses in space beyond the Earth's shadow, and there raising crops to support a population of emigres from the Earth. His astronauts visited the Moon, but only as an excursion in passing; their most important destination was the asteroids, a vast resource of materials.[1-6]

Still other authors, most of them writing later in the twentieth century, played with the idea of habitats in space. Lasswitz in 1897, Bernal, Oberth, Von Pirquet, and Noordung in the 1920's, continued the theme,[7-14] as did Wernher von Braun, Dandridge Cole, and Krafft Ehricke in the 1950's and 1960's.[15-26] Although many of these ideas are echoed in this book, it would have been difficult, before the year 1969, to make of them a coherent picture without serious technical gaps.

Our goal is to find ways in which all of humanity can share in the benefits that have come from the rapid expansion of human knowledge, and yet prevent the material aspects of that expansion from fouling the worldwide nest in which we live. Necessarily, many of the concerns of this book are materialistic, but more than material survival is at stake. The most soaring achievements of mankind in the arts, music and literature could never have occurred without a certain amount of leisure and wealth; we should not be ashamed to search for ways in which all of humankind can enjoy that wealth.

A firm schedule for the development of resources in space would depend on decisions not yet made, but it appears that construction of a high-orbital facility could begin within seven to ten years using launch vehicles no more advanced than those of today, and that it could be completed within fifteen to twenty-five years.

Governmental interest in high-orbital manufacturing stems in part from calculations on its economics. These suggest that a community in space could supply large amounts of energy to the Earth, and that a private, perhaps multinational investment in a first space habitat could be returned several times over in profits.

Much of the public interest relates to the human prospect that thousands of people now alive may choose within the next decades to live and work on a new frontier in space. If the concept is realized as soon as is technically possible, something like the following "letter from space" might be written within the next twenty years.

Dear Brian and Nancy:

I can understand that you want to hear from someone who's working and living in space before deciding whether to make the commitment yourselves.

According to your letter you've reached the "finals" in the selection process now. The next step will be the admission interview. After that, if you get an offer, you'll have to decide whether to go for the six-months' training. Though I never was in the Peace Corps, I understand that the selection methods are similar to theirs. Most people in your training group will pass the tests.

Then there's the big step of the first spaceflight, the three-week stay in orbit. By now the flight itself is quite routine; you'll find that the single-stage shuttle interior is much like that of one of the smaller commercial jets; there'll be one hundred and fifty of you traveling together. The g-forces will be higher than in commercial aviation, but still nothing to worry you. The trip into orbit will only take about twenty minutes, and then you'll experience something really new: zero-gravity. You may feel queasy at first—as if you were on a ship at sea. The three-week trial period is to sort out cases of severe space sickness, and to find out whether you are among those who can adapt to commuting each day between normal gravity and zero. That's important because our homes are in gravity obtained by rotation, and many of us work in the con-

Approaching Island One. Mirrors
collect sunlight for farms and
shielded living areas. Foreground
panels radiate waste heat.

struction industry, with no gravity at all. Those who can adapt to rapid change qualify for higher paying jobs. The trial period also gives people the chance to decide that "this is not for me."

After three weeks you'll be ready to transfer to one of the "liners" on its next trip in. Jenny and I enjoyed that voyage. You'll be on the Goddard or the Tsiolkowsky and each takes a week for the outbound passage. About half of the passengers will be newcomers like yourself, and half will be returnees coming back from vacations on Earth. The ship rotates, so there will be gravity, normal in the public rooms and less than that in the sleeping cabins. In the six-months' training period you'll have had cram courses in foreign languages, so try talking with some of your fellow passengers from other countries. We like visiting nearby communities for dinners out fairly often, and enjoy talking with people we meet there even though our foreign-language ability is mainly the "restaurant" variety.

In space near the communities, the biggest things you'll see will be solar satellite power stations being assembled to supply energy for Earth. Those power stations are about ten times as big as the habitats themselves. You won't see much detail from the outside of the habitats because they're shielded against cosmic rays, solar flares, and meteoroids by a thick layer of material, mainly slag from the processing industries.

All the habitats are variations of basic sphere, cylinder or ring shapes. We live in Bernal Alpha, a sphere about five hundred meters in diameter, with a circumference inside, at its "equator" of nearly a mile. We have track races and bicycle races that use the ring pathway. That path wanders all the way round, generally following the equator, and near it is our little river. Bernal Alpha rotates once every thirty-two seconds, so there is Earth gravity at the equator. The land forms a big curving valley, rising from the equator to 45 degree "lines of latitude" on each side. The land area is mainly in the form of low-rise, terraced apartments, shopping walkways, and small parks. Many services, light industries, and shops are located underground or in a central low-gravity sphere, or are steeply terraced, because we like to preserve most of our land area for grass and parks. Our sunshine comes in at an angle near 45 degrees, rather like mid-morning or mid-afternoon on Earth; the day-length and therefore the climate are set by our choice of when to admit sunlight. We keep Canaveral time, but two other communities near us are on different time zones. All the communities serve the same industries, so the production operations run twenty-four hours a day, three shifts, but with no one having to work the night shift.

Alpha has a Hawaiian climate, so we lead an indoor-outdoor life all year. Our apartment is about the same size as our old house on Earth, and it has a garden. Alpha was one of the first habitats to be built, so our trees have had time to grow to a good size.

You'll notice immediately the small scale of things, but for a town of 10,000 people we're in rather good shape for entertainment: four small cinemas, quite a few good small restaurants, and many amateur theatrical and musical groups. It takes only a few minutes to travel over to the neighboring communities, so we visit them often for movies, concerts, or just a change in climate. There are ballet productions on the big stage out in the low-gravity recreational complex that serves all the residents of our region of space. Ballet in $\frac{1}{10}$ gravity is beautiful to watch: dreamlike, and very graceful. You've seen it on TV, but the reality is even better. Of course, right here in Alpha we have our own low-gravity swimming pools, and our club rooms for human-powered flight. Quite often Jenny and I climb the path to the "North Pole" and pedal out along the zero-gravity axis of the sphere for half an hour or so, especially after sunset, when we can see the soft lights from the pathways below.

You asked about our government, and that varies a great deal from one community to another. Legally, all communities are under the jurisdiction of the Energy Satellites Corporation (ENSAT) which was set up as a multinational profit-making consortium under U.N. treaties. ENSAT keeps us on a fairly loose rein as long as productivity and profits remain high—I don't think they want another Boston Tea Party. There are almost as many different kinds of local government as there are national groups within the colonies; ours happens to be a town meeting style. That wouldn't work in a town of as many as 10,000 people, except for the fact that all of us are much too busy to make a hobby of electioneering,

7

and that the basics of habitat survival require a high level of competence on the part of the maintenance people. Our teenagers have to work a year in one of the life-support maintenance crews—it's a little like military service on Earth—and if the regular government or maintenance people were to get balky, they'd be replaced by volunteers awfully fast.

Jenny and I laughed a bit about your comment on having to give talks to civic groups—I remember we went through the same things ourselves.

For information to use in your lectures, I'll mention a few basics. The initial stock of water for each habitat is obtained by combining hydrogen brought from Earth with eight times its weight in lunar oxygen. Here at L5, oxygen is a waste product from the industrial processes that turn out metals and glass. Our soil, of course, comes from the Moon and is fertile once we add water and nitrates. Because of our unlimited cheap energy, we don't have pollution here. Where energy costs almost nothing, and raw materials are relatively expensive, it pays to break down every waste product into its constituent elements.

So far there aren't enough communities to make long-distance travel a problem, but when there are many of them, spaced over thousands of miles, we already know how the transport system will work. We can just accelerate an engineless vehicle to a high cruising speed by an electric motor at one community, and then after a trip of several thousand miles, we can slow it to a halt by an arresting cable at another community.

A long time ago someone calculated the maximum size for space habitats. They could be made in sizes at least as large as twelve miles in diameter, with a land area of several hundred square miles in each one. We're already talking about shifting the mining base from the Moon to the asteroids, where we'll have a complete range of elements including carbon, nitrogen, and hydrogen. In energy, it won't be any harder for us to get materials from the asteroids than from Earth, and it should be a lot cheaper because the transport system can take its time and won't ever need high thrust. Someone calculated how much "room for growth" there will be once we start to use the asteroidal material. The answer came out absurdly high: with the known unused materials

8

out there, we could build space communities with a total land area 3,000 times that of Earth.

To go on with our situation, it's a comfortable life here. Fresh vegetables and fruit are in season all the time, because there are agricultural cylinders for each month of the year, each with its own day-length. We grow avocados and papayas in our own garden and never need to use insecticide sprays. Of course we like being able to get a suntan without ever being bitten by a mosquito. To be free of those pests, it's worth it to go through the inspections before getting aboard the shuttle from Earth.

You asked whether we feel isolated. Some of us do get "island fever" to some degree, probably because we're really first-generation immigrants; it never seems to bother the kids who were born here. When you sign your contract there are clauses that help quite a bit though. One is the provision for free telephone and videophone time to Earth. Another sets up free transportation to Earth and return on a space-available basis. Jenny and I took six-months' leave after our first three years here. Our visit was luxurious, because our salaries are paid in part in Earth currency; we're both employed, Jenny as a turbine-blade inspector and I in precision assembly. Our housing, food, clothing, and the rest are purchased in SHARES (Standard High-orbital Acquisition-units Recorded Electronically) so our Earth salaries just accumulate in the bank. When we went back we had a lot of money to spend, and even on a luxury basis we couldn't go through it in six months.

We found something though, that may help to answer your basic question: by the time the vacation was nearly over, we were very ready to come back here. We missed our own place. Jenny is an enthusiastic gardener, and though other people were living in our apartment here and taking care of the greenery, she wanted to be at home to enjoy it herself. And I missed the friends I'd been working with. I can best describe the other thing that drew us back by saying that the space habitats are exciting places to be. They're growing and changing so fast that if you're away for six months you've missed a lot.

As to whether you'll really like it, of the people who came with us, more than half intend to stay after their five-year

9

contract is up. I understand that the settlement of Alaska has had about the same kind of "stay-ratio."

Now we're beginning to ask ourselves: will we want to retire to Earth or not? We don't have to face that for another twenty years, but we can see already that it won't be an easy decision. Some of us who are handy with tools have formed a club to design spacecraft for our own construction—rather like the homebuilt-aircraft clubs on Earth. We're thinking of homesteading one of the smaller asteroids, and the numbers look reasonable. Especially if our daughter and son-in-law decide to come along, with the grandchildren, I think we're more likely to move further out than go back.

If you decide to come out, let us know what flight you'll be on and we'll meet you at the docks. We'd like you to come over to our place for supper, and we'd be glad to help you to get settled.

With our best wishes for good luck on the tests,
Cordially,
Edward and Jenny

As we explore these possibilities we must remember that they are just that—not predictions or prophecies. The time scale may be longer than the fifteen to twenty-five years I estimate to be an achievable minimum; or I may be too cautious, and events may dictate a still faster scale. The "when" is not science but a complicated, unpredictable interplay of current events, politics, individual personalities, technology and chance. As a guess, though, I consider it unlikely that the first community in space will be established in less than fifteen years, and also unlikely that it will be delayed for another fifteen years beyond that. Neither of these dates is very far off; both are within the life-span of most people now alive. In the matter of dates, it is to me rather thought-provoking that Konstantin Tsiolkowsky, the great visionary space pioneer of nineteenth-century Russia, was himself too conservative on the date of the first Earth-orbital flight: he guessed the year 2017.

Robert Goddard (1882-1945), much of whose life was spent in the more practical and therefore much more difficult task of reducing the theory of rocketry to working hardware, left

Transfer ship docks at Island
One. Arriving passengers will float
through transfer corridor, then
walk down to normal gravity.

us with a caution lest our vision be too narrow:

"It is difficult to say what is impossible, for the dream of yesterday is the hope of today and the reality of tomorrow."

2

THE HUMAN PROSPECT ON PLANET EARTH

We now have the technological ability to set up large human communities in space: communities in which manufacturing, farming, and all other human activities could be carried out. Substantial benefits, both immediate and long term, can accrue to us from a program of expansion into that new frontier.

13

The normal first reaction to such a statement is disbelief: isn't such a development beyond us? Not at all: the settlement of space by humans could be carried out without ever exceeding the limits of the technology of this decade. But even if it is possible, should we make the effort? I believe we should; the reasons go from an immediate and severely practical one: solving the energy crisis which we face here on Earth; to the slightly longer-term problem of population size and Earth's capacity to support it; finally to a nonmaterial problem, compelling but not to be reckoned in dollars: the opportunity for increased human options and diversity of development.

Through many tens of thousands of years human beings were few in numbers, and insignificant in power over the physical environment. Not only war but famine and plague decimated populations whenever they grew large; centuries passed without great increase in the total human population. The quality of life, for most people in those preindustrial years, seems to have been low even in times of peace. Although there were, nearly everywhere, small privileged classes enjoying comparative wealth, most people lived out their lives in heavy labor, many as slaves. [1] Through all that time any observer on another planet would have found it very difficult to find telescopic evidence for the existence of the human race; our power over Earth was too slight to be noticeable.

Very suddenly, in a time of less than two hundred years, our human status as passengers on a giant planet, lost in its immensity and powerless before its forces, has changed dramatically. The beginnings of a science of medicine, and the rapid development of chemistry, have made fatal disease a rarity among children in the wealthy nations, and have even reduced its power in the poorer nations. With that one radical change we suddenly find ourselves growing in numbers so fast that Earth itself cannot long sustain our increase.

At the same time, our power to change the surface of Earth has increased: our activities can and now do alter the planet and its atmosphere. We achieve every year a greater degree of control over the natural environment, and we change it more in attempts to suit our liking. The result, though, does not always please us.

The industrial revolution has been the mechanism by which our physical power has increased, and by which, for the first time, a substantial fraction of the human population has reached a high living standard. Comfort, a reasonable life-expectancy, freedom to travel, the easy availability of news and education—these have come to the most advanced countries as benefits of industrialization. But that process has brought evils as well. Though it began only two hundred years ago, less than a ten-millionth of the time since Earth was formed, its side effects have already altered Earth in frightening ways. It has scarred, gutted, and dirtied our planet to a degree many people find intolerable. Smoke and ash from factories in England cloud the air as far away as Norway; pollutants from the industries of Japan can be detected in the snows of Alaska. Nearly every major city has its air-pollution problem.

If those evils had occurred after the industrial revolution had penetrated to every nation on Earth, we could have tried to discuss, as a species, the actions necessary to counteract them. We are not so fortunate; the evils of environmental damage are minor compared to others that have appeared: sharp limits on food, energy, and materials confront us at a time when most of the human race is still poor, and when much of it is on the edge of starvation. We cannot solve that problem by a retreat to a pastoral, machine-free society: there are too many of us to be supported by preindustrial agriculture. In the wealthier areas of the world, we depend on mechanized farming to produce great quantities of food with relatively little human effort; but in much of the world, only backbreaking labor through every daylight hour yields enough food for bare survival. About two-thirds of the human population is in underdeveloped countries. In those nations only a fifth of the people are adequately fed, while another fifth are "only" undernourished—all the rest suffer from malnutrition in various forms.[2]

In those countries the need to increase the food supply is desperate. When the land cannot support its population, and starvation is general, disease strikes at the old and even harder at the young. Small children of a family contract the crippling diseases of malnutrition; parents must watch their children die, and be powerless to save them. In such areas some degree of industrialization is not a luxury but a desperate need; it is a great tragedy of the late twentieth century that the satisfaction of such a need is being denied or delayed in part because of the energy and materials limits of Earth.

As we view the process which has given to most people in the industrialized world some freedom of movement and relief from heavy labor, we find that it is based on the increasing use of artificial energy sources. Within our own lifetimes we have seen rapid long-distance travel become commonplace for a large fraction of the population; forty years ago it was impossible even for the very rich. A luxury passenger liner of the 1930s took several days to cross the Atlantic, and its engines developed about twenty horsepower per passenger carried. Now, a crossing by jet aircraft takes only a few hours, but the plane needs several hundred horsepower per passenger. Until the energy crisis of 1973–74, energy usage in the United States was growing by 7 percent per year.[3] The mechanization of agriculture, the "green revolution," and the rapid development of nonfarming industry in the emerging nations all depend on their going through a similar period of rapid growth.

They are having a hard time doing so: in energy usage, we were there first, and have skimmed the cream of the Earth's easily available energy sources.

From a political and moral viewpoint, we in the developed nations bear a responsibility for the plunder of the past centuries. It is unlikely, though, that a large segment of the population in the advanced countries is going to reduce its living standard by a substantial amount, voluntarily, in order to share the energy wealth of Earth with the emerging nations. As I will show, there may be an acceptable alternative: a way in which inexpensive, inexhaustible energy sources can be made available to the developing nations without self-denial on our part.

Any technological solutions we employ to solve our

problems must, though, retain their logic over a very long time-span. As E. F. Schumacher put it:

"Nothing makes sense unless its continuance for a long time can be projected without running into absurdities ... there cannot be unlimited, generalized growth ... Ever bigger machines, entailing ever bigger concentrations of economic power and exerting ever greater violence against the environment do not represent progress: they are a denial of wisdom."[4]

These considerations should be in our minds as we examine the technical suggestions contained in this book. I would put them in the form of guiding principles:

1. A proposal to improve the human condition makes sense only if, in the long term, it has the potential to give all people, whatever their place of birth, access to the energy and materials needed for their progress.

2. A technical "improvement" is more likely to be beneficial if it reduces rather than increases the concentration of power and control.

3. Improvements are of value if they tend to reduce the scale of cities, industries, and economic systems to small size, so that bureaucracies become less important and direct human contact becomes more easy and effective.

4. A worthwhile line of technical development must have a useful lifetime "without running into absurdities" of at least several hundred years.

There are other needs which should, I believe, be met by any development of our industrial society, if it is to be successful. It would be desirable if the noise and pollution of our transportation systems could be removed from the environments in which we have our homes and raise our children. Yet we must preserve the freedom of rapid motion even to great distances.

We should also strive for a solution to the problem of unwanted growth in our individual environments; if population growth continues, we should look for a way in which it can do so while still allowing each individual human town to be stable in size and density.

Finally, as we strive to find solutions to the physical problems faced by mankind, we must realize, with humility, that we can offer no panaceas. There are no Utopias.

Mankind does not change, and retains always the capacity for evil as well as for good. At the most we can suggest opportunities whose technical imperatives will make it easier for mankind to choose peace rather than war; diversity rather than repression; human simplicity rather than inhuman mechanization. Technology must be our slave, and not the reverse.

Within the past decade four problems have been recognized, all of which relate to the limited size of Earth: they are energy, food, living space, and population. The last of these is basic to the other three; therefore we must know the predictions for growth in the human population, and should estimate the accuracy of those predictions.

The basic sourcebooks for demography are the publications of the United Nations Department of Economic and Social Affairs. There have been four attempts over the past twenty years, by that Department, to summarize worldwide statistics and predict world population growth. The last of these was published in 1973.[5] The resources that were employed for that study are probably at least as great as those available to any other scholarly group.

As a starting point, two numbers are well known: the present world population (just over four billion people, that is four thousand million) and the population-growth rate. For several years that last number has averaged 2 percent annually, corresponding to a doubling time of thirty-five years for the world population.

Viewed on a time scale of many centuries, though, the population-growth rate has itself increased continuously. This has led to such papers as that of Von Hoerner, which shows that up to 1970 the best mathematical fit to the population-growth curve would lead to a true "explosion": an infinite number of people about fifty years from now.[6] This sort of study is of great value in calling attention to the growth problem, but it is best understood as a statement that within the next few decades the growth rate must reduce, and radically. For purposes of this book I will use the much more conservative growth-rate figures of the U.N.: the situation is already serious enough without the need to overstate it.

The total world population in 1980 has been estimated by the U.N. Department of Economic and Social Affairs in each of its four summaries, beginning in the early 1950s. Significantly, in each successive revision the Department has raised its estimate of the population of the world in the year 1980. As that date grows closer and the extrapolations can be based on more accurate information, the Department has found that its previous estimates have been too low.

We must also assess the kind of biases that may be put into the Department's numbers as a result of inevitable political pressures. During the past few years many nations have introduced population-control measures, enforced either by economic bribery (as in India, where the payment to a young man for undergoing an irreversible vasectomy is typically a quarter of a year's salary) or by social and governmental pressure (as in China, where early marriage is forbidden and where a third child is barred from receiving governmental welfare benefits). When a United Nations member state tells the Department that it has such a program in force, the Department can do little but take the statement at face value. Its predictions, therefore, generally reflect the assumption that the population-control programs will be successful as planned. The risks contained in that assumption were illustrated in 1977, when population control became a political issue in India, and the government which had attempted such control was overthrown at the polls after being in office for many years. Even with a successful population-control program in the underdeveloped nations, the Department tells us that there will be about six and a half billion people in the year 2000. Growth over this last quarter of the twentieth century within the developed nations will be slow; the increase will come almost entirely within the poor nations. South and East Asia alone will have, by the year 2000, more people than the entire world had in 1970. On the average the one-third of the human population that now lives in developed nations is adequately well-off in medical care, education, food, and material possessions, though many of the larger developed nations have serious problems of internal inequities. By the end of the century, though, an even smaller fraction of all people will live in developed nations, according to the United Nations. The world of 2000, then, will be poorer and hungrier than the world of today.

This growth in population seems contradictory, if we recall that the Department's numbers reflect an optimistic assumption about population-control programs. There is no contradiction, though: the anticipated increase in numbers will be the result of a distorted age-structure of the populations in the poor nations. There, medical advances have come so recently that now most of the population is very young, well below childbearing age. Even if those young people have only two children per couple, the populations of their countries will rise greatly over the next generation.

Knowing that fact, we must also recognize that for the poorer nations *not* to experience rapid population growth over the next twenty-five years, they would have to adopt violent measures. It would not be enough to limit family size to two children; it would be necessary for those nations to suppress new births to levels probably unachievable except by massive, forced sterilization.

The U.N. studies assumed that population-growth rates would fall toward the end of the century. The U. N. hardly dares to predict what will happen beyond that time, but if we project their graphs we find that the 10 billion mark will be reached by 2035. Most of the "new people" will be in the underdeveloped nations, and will be born into poverty. And remember—that's the "good" news, based on the idea that population-control programs will succeed.

By the same token, as time goes on the U. S. fraction of the total world population will become more and more insignificant. By the turn of the century only one human being in twenty-five will be American, and only one in fifty of the new births will be in this country. As far as the total world situation is concerned, therefore, it hardly matters what happens to our own low birth rate.

Though I've used the U.N. figures in estimating how the world population will grow during the next few decades, there are two reasons for being uneasy about doing so: the U.N. figures are based on an assumption that growth rates in the poor nations will be reduced drastically, partly as a result of industrialization. There are, though, serious barriers to the industrial revolution in those nations. Second, the U.N. has been too conservative in its previous estimates; it may be so again. Third, to achieve a downturn in population-growth rate will mean reversing a trend[7] which has existed for at least 2,000 years. That may not be easy.[8]

In the rich countries, the comfort, the abundance, and the freedom of choice enjoyed by most people are achieved only by a high rate of energy use. We grow food efficiently only by spending energy to make chemical fertilizers;[9] our houses are lighted, powered, heated, or cooled at the expense of energy; our freedom to travel depends on burning, every year, an amount of fuel which is many times our own weight.

In the United States we now use energy in all its forms at a total rate of about 10,000 watts per person; until the energy crisis of 1973–74, that use rate was doubling every eight years. Not all of that expenditure of energy is necessary, but our experience during the 1974 gasoline restrictions taught us that not much saving in energy can be made without a noticeable reduction in each individual's freedom of movement. If energy shortages are going to become chronic, we must not forget what they will mean, not only in terms of our inconvenience, but in terms of sheer survival within the poor nations. We must also recognize that conservation is only a palliative; we will continue to need new sources of energy.

At present we in the United States are sharply aware of the need for energy conservation. A number of energy-saving schemes have been tried already, but the people whose business it is to anticipate future use of power predict, at best, a lower rate of increase than was common up to 1974.[10,11]

In the U. S. we now burn about half a billion tons of oil products every year, and our total energy usage is about two and a half times as great.[12] A rise in the living standard of the underdeveloped nations to our level would require a similar use of energy.

If the entire population of Earth were to be using energy

at the same rate as we do, and were obtaining it from the same mix of oil, coal, gas, and other sources, the world total of proven oil resources would be used up in about four years. Even with a strong program of conservation, our use of energy would still be so high that if the whole world were to be at our standard of living, and getting all its energy from oil, by the turn of the century the world use rate would burn up the world's proven resources in six months.

There are, of course, large quantities of oil not yet on the "proven reserves" list, but their recovery will probably be at a cost to the environment. In the United States, where the environmental movement began and is strong, there is already concern about the cost to our natural environment of exploiting lowgrade, remote and undersea sources: for oil, it means the ugliness of the drill rigs in the Santa Barbara Channel, and the dangers of the controversial Alaskan pipeline. For coal or oil shale, it means strip-mining. For nuclear fuels, it means mining and crushing surface rocks over large areas of the Western mountain landscape.

Inexpensive, abundant sources of energy have been the basis of the industrial revolution so far. Now, when energy costs are rising sharply, those costs may well be contributing to inflation and economic stagnation in the "wealthy" nations. In a single year, 1973–74, the world price of crude oil quadrupled.[13] That single increase cost our U.S. economy more than twenty billion dollars for every year that followed.

In poor, heavily populated countries rising energy costs are even more serious: in order to grow enough food to lift their increasing populations above the starvation level, those countries must convert very rapidly to intensive agriculture. This conversion will require greatly increased fertilizer production, and that in turn will demand energy.[14]

So far, nuclear power has provided only a small fraction of our energy needs. As fossil fuels grow scarcer and more expensive, most experts think we will be forced to rely more and more heavily on nuclear fuels. The prospect is not an attractive one: the study prepared by Associated

Universities, Inc., foresaw most of our electric power coming from liquid-metal fast-breeder reactors within three decades.[15] The problem of the disposal of their radioactive wastes would not be easy to solve. In addition, these reactors would produce plutonium, from which atomic bombs in large numbers could easily be made. It seems likely that in that case nearly every nation, whatever its size or political stability, would have its stock of nuclear weapons. Large quantities of fissionable materials would be shipped about, and almost inevitably some would also be hijacked by terrorist groups.[16]

For many years we have looked toward nuclear fusion as a clean power source; but even after twenty years of effort and billions of dollars of investment in research no laboratory has succeeded in achieving it. As development has gone on it has also become clear that nuclear fusion will not be so clean a source as originally hoped; it too will produce substantial radioactive wastes. I do not consider fusion-power research a waste of time, but it is important to realize that fusion power would require a technology far more difficult, advanced and speculative than anything suggested in this book.

Solar energy would be a good solution to our energy problems, if it were available twenty-four hours per day and were never cut off by clouds. We should not dismiss it entirely, but it is very difficult to obtain at Earth's surface when we need it. To summarize, our hopes for improvement of the standard of living in our own country, and for the spread of wealth to underdeveloped nations, depend on our finding a cheap, inexhaustible, universally available energy source. If we continue to care about the environment in which we live, that energy source should be pollution-free and should be obtainable without stripping Earth.

It could be argued that in the most developed countries a slowing of the growth rate in energy usage could occur without serious hardship; that may be true, although I have the uneasy feeling that there may be a connection between energy shortages, price rises, and the present serious economic problems in all the industrialized, energy-consuming nations. In the underdeveloped nations, for which the industrial revolution is still to occur, rapid growth rates in energy usage are probably a condition of development. For a healthy world economy it may therefore be necessary to assume that the growth rates which have existed up to now (about 7 percent per year in energy) will have to continue. It has been pointed out by Von Hoerner that if such growth continues, within about eighty-five years the power we will be putting into the biosphere will be enough to raise the average temperature of Earth's surface by one degree centigrade.[17] That is enough to cause profound changes in climate, rainfall, and in the water level of the oceans. Some geologists feel that the ice ages of the past were brought on by temperature changes no larger than that.

I think Von Hoerner is basically right. We can make our own independent estimates, and come up with similar results. Using the "optimistic" low growth rate in population projected by the United Nations, by the year 2060 there will be some 13 billion people. If at that time the present great disparities in the wealth of nations have been reduced, so that all are using energy at about the same per capita rate, that maximum tolerable rate turns out to be greater than our own by an amount that is only 3 percent per year of per capita growth. The "heat limit" is therefore a real one. It may be that it could be pushed back, for a while, by covering large areas of Earth with mirrors to reduce the total of absorbed solar energy. But it cannot be delayed for long: another fifty-five years and we would be putting into the biosphere ten percent as much heat as is received by the Sun. A continual growth of energy usage on the surface of Earth, therefore, even if the growth rate is moderate, is one of the "absurdities" of which Schumacher has written.[18]

Professor Robert Heilbroner has studied the consequences, for human political and social development, of the energy and materials limits we have just discussed.[19] He assumes, in my opinion rightly, that people will continue to be guided by the same desires, instincts, and fears that have dominated human history up to this point.

He dismisses the notion of arresting the industrial revolution at its present level: ". . . impassioned polemics against growth are exercises in futility today. Worse, they may even point in the wrong direction . . . In the backward areas, the acute misery that is the potential source of so much international disruption can be remedied only to the extent that rapid improvements are introduced, including . . . health services, education, transportation, fertilizer production and the like."

He is pessimistic about the prospects for widespread social change either within the capitalistic or socialist systems: "We have become aware that rationality has its limits with regard to the engineering of social change, and that those limits are much narrower than we had thought . . . that growth does not bring about certain desired ends or arrest certain undesired trends." In his opinion, as a result of the increasing scarcity of energy and materials, ". . . a climate of extreme 'goods hunger' seems likely to result. In such a climate, a large-scale reorganization of social shares would have to take place in the worst possible atmosphere, as each person sought to protect his place in a contracting economic world."

Under these conditions Heilbroner feels that the threat of nuclear war is likely to increase greatly in the next decades; because of energy and materials limits, ". . . massive human deterioration in the backward areas can be avoided only by a redistribution of the world's output and energies on a scale immensely larger than anything that has hitherto been seriously contemplated . . . such an unprecedented international transfer seems impossible to imagine except under some kind of threat.

"Yet two considerations give a new credibility to nuclear terrorism: nuclear weaponry for the first time makes such action possible, and 'wars of redistribution' may be the only way by which the poor nations can hope to remedy their condition."

Even if nuclear war does not occur, and humanity staggers on for another two or three generations, Heilbroner feels that the heat emission limit poses: ". . . a challenge of equal magnitude for industrial socialism as for capitalism—the challenge of drastically curtailing, perhaps even dismantling, the mode of production that has been the most cherished achievement of both systems. Moreover, that mode of production must be abandoned in a mere flash of time as historic sequences are measured."

Heilbroner points out that even in the decades immediately ahead we will be forced to turn to increasingly authoritarian governments: "... the passage through the gantlet ahead may be possible only under governments capable of rallying obedience far more effectively than would be possible in a democratic setting." "... strong leaders provide a sense of psychological well-being that weak ones do not, so that in moments of crisis and strain demands arise for the exercise of strong-arm rule." He concludes that intellectual freedom of expression is almost sure to be sacrificed to the exigencies of the energy and materials limits: "... suppose ... that only an authoritarian, or possibly only a revolutionary, regime will be capable of mounting the immense task of social reorganization needed to escape catastrophe ... might not the people of such a threatened society look upon the 'self-indulgence' of unfettered intellectual expression ... as of no concern, or even of actual disservice, to the vast majority?"

There is, of course, an alternative to industrial growth. Conceivably, perhaps after a series of catastrophes, mankind would adapt a static society. This alternative, a "steady-state" civilization, was considered by J. W. Forrester, leader of the M. I. T. systems analysis team which produced, with the support of the Club of Rome, the document "Limits to Growth."[20] By calling attention to the consequences of exponential growth in a finite environment, that group performed, in my opinion, a great service. Detailed shortcomings of the computer model used are unimportant by comparison. Forrester could see no viable alternative but a rapid switchover of our present civilization to a steady-state mode. Heilbroner comes to a similar conclusion: "In our discovery of 'primitive' cultures, living out their timeless histories, we may have found the single most important object lesson for future man."

A steady-state world order need not be primitive; for

26

example, the pre-Conquest world of the Inca in Peru was a rigidly structured, dictatorial society satisfying a steady-state condition. A peasant of the Inca empire went through life with all his duties and responsibilities rigidly specified, and at his death left a world almost exactly the same as the one he was born into. Almost any static society is forced in self-defense to suppress new ideas. In Heilbroner's words: "The search for scientific knowledge, the delight in intellectual heresy, the freedom to order one's life as one pleases, are not likely to be easily contained within the tradition-oriented static society. . . ."

Professor Heilbroner is frank to admit that ". . . many conclusions in this book have caused great pain to myself . . . the human prospect, as I have come to see it, is not one that accords with my own preferences and interests, as best I know them." And finally, "If then, by the question 'Is there hope for man?' we ask whether it is possible to meet the challenges of the future without the payment of a fearful price, the answer must be: No, there is no such hope."

3
THE PLANETARY HANGUP

The exponential growth of population, on what has become not only a finite but now a sharply limited planet, is almost certain to make the decades ahead on Earth very difficult, and perhaps catastrophic. In the United States, even cushioned as it is by previous wealth, we are feeling the pinch of unemployment, rapid inflation, and conflict between industrial efficiency and environmental protection.

If we look in detail at the population-growth rates of individual countries, we find that stability has been reached in those areas which have achieved wealth at a high technological level by intense use of energy: North America, Europe, and Japan. To maintain that growth of wealth, these countries must burn fossil fuel reserves at a frightening rate. Between the Persian Gulf and Japan there is a continuous chain of oil tankers, spaced so closely that the crew of one can see the smoke of the next.[1] Our own U. S. appetite for fossil fuels is even greater.

In past centuries plagues and wars were an important factor in holding populations stable. Where poverty is widespread and improvements come slowly, as in South America, Africa, and India, population-growth rates remain explosive. Poverty and ignorance go hand in hand, and the decision to limit family size can best be made by families freed of heavy manual labor, secure in good health care for their children, and wealthy enough to spare their children's time from the fields for education.

It seems then that the key to a low population-growth rate may be wealth; conversely, we must be somewhat apprehensive about the accuracy of the U. N. estimates on population-growth rate: those estimates, even though they predict a population three times its present size within one human lifetime from now, are based on the assumption that growth rates in the poor nations will drop sharply long before that. If, though, there is no way that the underdeveloped nations can become significantly wealthier, the corresponding drop in population-growth rates may not come except by catastrophe.

If we want population-growth rates to drop by peaceful means, it seems that the best way may be to attack the issues of poverty and ignorance: we need to increase the wealth of the underdeveloped countries not just by a few percent per year but massively, by factors of ten or a hundred. We cannot begin to do this by giveaway programs; we do not possess the enormous amount of wealth needed, and the historical evidence seems to be that small efforts to help are usually cancelled by population increases. The areas of the world with the worst problems are often energy-poor or located in miserable climates, so their long-term prospects for industrialization don't justify optimism.

Somehow we must find a way to bypass those limits, and to set up a chain reaction in the production of new wealth; a reaction that we may trigger but which must then sustain itself as it grows. It will be of little value unless the doubling time for wealth is quick compared to the population doubling time in the poorest areas—and that means short compared to eighteen years.

We are in a period in which technical change comes rapidly; often the results of change are mixed or heavily adverse; yet we cannot stand still: to do nothing is itself an action, for it is to condemn millions in our crowded world to certain death by starvation. What actions can we take that will reverse the present trend toward increasing poverty and hunger?

Several years ago Gerald Feinberg addressed the issue of technical change in a book subtitled "Mankind's Search for Long-Range Goals."[2] I would take issue with Feinberg on only two points: we do not usually "search" for goals; most people have enough to do to cope with their own lives, and leave the long-range or large-scale questions to chance, with perhaps a dim hope that "something will turn up." Second, it was Feinberg's suggestion that major issues be submitted to as large a fraction of the world's population as possible. He had in mind particularly such potentially explosive issues as artificial genetic change, the alteration of personality by chemicals, and increased human longevity. The idea that large issues should be debated by many people, not just a power elite, is in my opinion a good one: in the United States during the past decade it has been put into practice with good effect by voluntary citizen movements in the areas of family planning, environmental protection, and land conservation. We must recognize, though, that a population must be relatively wealthy, well-educated, and have considerable leisure if it is to spare time and effort for such debates. In the areas of our world where the problems are most severe,

31

almost no one can spare the effort to think beyond the next meal.

This is one of the rare occasions in human history in which a new technological option is being subjected, deliberately, to wide popular debate before, not after, the decision to go ahead with it has been made. I prefer it that way: I believe that the concept of the humanization of space can stand on its own merits, survive detailed numerical checks, and survive logical debate; to support it requires no act of faith, only the willingness to study unfamiliar ideas with an open mind. In keeping with Feinberg's strictures, in my opinion the long-term goals we should set, relevant to space habitation, should only be those with which nearly every rational human being, possessed of goodwill toward others, could agree. I think that the following goals satisfy that criterion, and that they should be our most important goals not only for humanitarian reasons, but for our own self-interest; and I do not believe that those two justifications must necessarily be in conflict.

1. Ending hunger and poverty for all human beings.

2. Finding high-quality living space for a world population which will double within forty years, and triple within another thirty, even if optimistic estimates of low-growth rate are realized.

3. Achieving population control without war, famine, dictatorship, or coercion.

4. Increasing individual freedom and the range of options available to every human being.

We in this country are certainly going to become increasingly insignificant as the years go on, both because of our decreasing relative numbers (only 4 percent of the world's population by the year 2000) and because of the energy and materials limits to the growth of our wealth. Is it reasonable then to set a fifth, more parochial goal? Realizing our limitations, should we not seek a role for this country that can be of benefit to humanity as a whole, and at the same time can benefit directly our own people and our own economy?

Considering the first four goals in the context of the fifth, it should be clear to us that we have no special magic to export in regard to governmental systems. Most of us are passionately attached to a democratic form of government, but it does not travel well: much of the world has rushed to imitate our technology and our systems of productivity; at the same time, there has been no such rush to imitate our system of government. We must also recognize that other systems have been found to work, perhaps not very much worse than our own, even in societies that have achieved industrialization. I own to a private belief that wealth and leisure, shared generally by a large segment of a population, are powerful forces tending toward more democratic forms of government, but I suspect that if the human race does achieve general affluence, and with it an increase in real human freedoms, it will do so within the outward forms of many different forms of government, and with many of the old polemic catchphrases still in regular use.

Can we point the way to an exponential growth of wealth, which could continue for many centuries, and which could be shared by all people? If we can, and can further lead the way to it by virtue of techniques in which we are acknowledged the leaders, we will have done something, as a nation, very worthwhile indeed, something far more worth looking back on with pride than lost dominance or a vanished empire. To achieve such an exponential growth of wealth, and therefore the opportunity to reach the four great goals listed, we would need:

1. Unlimited low-cost energy, available to everyone rather than just to those nations favored with large reserves of fossil or nuclear fuels.

2. Unlimited new lands, to provide living space of higher quality than that now possessed by most of the human race.

3. An unlimited materials source, available without stealing, or killing, or polluting.

Nothing in our solar system is truly unlimited, of course; no expansion can go on forever; but an exponential growth of wealth can be considered rationally if we can find the environment in which that growth can proceed for many hundreds of years; there is an enormous difference between sharp limits, forced on us within years or decades at

33

a time when most of us are still in deep poverty, and limits reached only after several hundreds or thousands of years, under conditions of high prosperity and universal education in a generally affluent and literate human population.

We are so used to living on a planetary surface that it is a wrench for us even to consider continuing our normal human activities in another location. If, however, the human race has now reached the technical capability to carry on some of its industrial activities in space, we should indulge in the mental exercise of "comparative planetology." We should ask, critically and with appeal to the numbers, whether the best site for a growing advancing industrial society is Earth, the Moon, Mars, some other

NASA concept for
early workbench
in space.

planet, or somewhere else entirely. Surprisingly, the answer will be inescapable: the best site is "somewhere else entirely."

In a roundtable TV interview, Isaac Asimov and I were asked why science-fiction writers have, almost without exception, failed to point us toward that development. Dr. Asimov's reply was a phrase he has now become fond of using: "Planetary Chauvinism."

What do we need for the exponential growth of wealth? Three things: energy, land area, and materials. The next question is: How much of them will we need, if growth of any sort is to go on? Suppose that a universally affluent, energy-rich, educated human population has about as low a growth rate as would be noticeably different from zero:

an increase of the total human population by about one-sixth over a human lifetime. That very modest rate of increase, considerably less even than now exists in the developed nations on Earth, would result in total growth by a factor of 20,000 over a period of 5,000 years. Right now, of course, growth is ten times as fast.

The conclusion we have to draw from those facts is that for exponential growth of wealth over a time-span long enough to make a real qualitative difference in human history, the factors needed in energy, land area, and materials are not just two or four or even ten: they are at least in the thousands, and probably in the hundreds of thousands. It is with that in mind that we must view Earth—and its "competitors" as a site for a large industrialized civilization.

The energy limits on planet Earth were discussed in the first chapter. Even if some inexhaustible source of energy is discovered and exploited here, we will reach the heat barrier in about one and a half human lifetimes; we cannot base an expanding industrial civilization on a site where a fundamental limit will be reached that soon.

The land-area resources of Earth are known; its geometry as a sphere in space determines that some of its areas are heated moderately, others too much or too little.

We could, in principle, make all the land area of Earth habitable, including Antarctica, and we could float colonies on the oceans; the resulting changes in worldwide climate would be profound, and there would be a severe risk of melting the ice caps and precipitating another ice age, but we would be forced in that direction if we had no alternative. The era in which virgin land of good quality in a good climate was available for settlement is long past; the United States is a relatively uncrowded country by world standards, but already our fastest growth is in regions (Arizona, New Mexico, and other desert areas) which would not attract large numbers of people if there were no air conditioning. In areas of California, once regarded as highly desirable, overcrowding has become so bad that in a recent survey about a third of the Californians said they'd rather be living in some other state. But the mood in nearby states of low population density (Oregon, Idaho, and others) is openly hostile to emigrants from California.

In Europe, the Netherlands is already near the saturation point for population, given its climate and growing season. In much of Asia the crowding of the land is still more serious, and it is there that the great population increases are still to come.

The prospects for colonization of other planetary surfaces are unappealing. First, the total areas involved are too small: the Moon and Mars total only about the land area of Earth; neither has an atmosphere. Both have the wrong gravity for maintenance of our bodies in good health, and the Moon has a fourteen-day night which would require any colonists there to do without natural sunshine for weeks at a time. Venus is an inferno hot enough to melt some metals, and would be uninhabitable without extensive "terraforming" of a kind well beyond our present capabilities. Even after such a conversion, it would still be unbearably hot, because of its location so much nearer the Sun than is our Earth. Finally, the total area of Venus, about equal to that of Earth, would be worth only two or three decades of time in terms of present growth rates.

Travel away from a planetary surface requires high thrusts and precise timing, and is therefore relatively difficult and expensive. On a planetary surface we are the "gravitationally disadvantaged," at the bottom of a deep hole in potential energy. From Earth, to raise ourselves into free space is equivalent in energy to climbing out of a hole 4,000 miles deep, a distance more than six hundred times the height of Mt. Everest. Does it make sense to climb with great effort out of one such hole, drift across a region rich in energy and materials, and then laboriously climb back down again into another hole, where both energy and matter are more difficult to get and to use?

There are still other disadvantages of basing an industrial civilization on a planetary surface:

Solar power: On Earth it is attenuated by the atmosphere, uncertain due to weather, and cut off every night by Earth's rotation. The average of solar energy input[3] to

37

the United States, over a year, is only about 0.18 kilowatts/m². In free space at a distance no farther than the Moon, but well away from both Earth and Moon, solar energy is available full time at a rate of 1.4 kilowatts/m²—that is, almost ten times higher than at Earth's surface, when averaged over a year; and it is never cut off by night.

Travel and shipping: On a planet with atmosphere both are slow, and wasteful in energy. In the U.S. transportation system, about a quarter of all the energy we spend goes into fighting gravity and atmospheric drag—that's a waste of around two and a half tons of petroleum each year for every man, woman, and child in our country.

Confinement to one gravity: Until the last decade, it would never have occurred to us to think that industry could operate in zero-gravity; but if that option is presented to us, it can be used to good advantage. Every activity involving massive objects or large weights of materials is dominated by the cranes, rails, engines, and other machinery needed to handle heavy objects in Earth-normal gravity. In zero-g, all that would be unnecessary. There are industrial processes, such as the growth of perfect large single crystals, which are impossible in one-g, but easy in zero-g. Single crystals can be ten or twenty times as strong for their size as the same materials in less ordered form.

Climate, the locations of materials, and the special property of oceans for cheap transportation tend to cause wide separations, here on Earth, between agricultural-producing areas and population centers. As a consequence, we become tied into interdependent networks thousands of miles in extent. Anyone who interrupts one of those networks by cutting off our sources of energy, of food, or of materials can hold a large population for ransom. We have examples of that sort of threat frequently, and the result is always the same: even at best, prices are driven up, production is slowed down, and almost everyone suffers. At worst—and that worst is approaching with frightening speed as we plunge deeper into the energy-food crisis—we approach a world society governed by mutual threat: deprive me of oil, and I deprive you of food;

threaten me enough with deprivation, and when I have nothing left to lose I will risk life itself in a last desperate gamble; provide for me, or I burn you to death with hydrogen bombs.

The same factors of climatic variation, the need for sea transport to minimize transport inefficiencies caused by gravity, and the seasonal cycle tend to produce very large concentrations of population—living in numbers so great that they are continuously subjected to the evils of bigness: high crime rates, dirt and disease, social alienation, and political corruption.

Up to now, we have taken it for granted that huge cities were an inevitable part of industrialization. But what if it were possible to arrange an environment in which agricultural products could be grown with high efficiency, anywhere, at all times of the year? An environment in which energy would be universally available, in unlimited quantities, at all times? In which transport would be as easy and cheap as ocean freight, not just to particular points but to everywhere? There is, now, a possibility of designing such an environment, and it will be the topic of the next chapter.

The decrease in human options: The solution to the problems of energy and materials would not guarantee freedom and well-being for all: we have had too many examples in history of man's capability for inhumanity for us to assume that. Until very recently, though, we had some hope that averaging over the ups and downs the human race as a whole was struggling toward more decent living conditions, better education, and more freedom. The ignorance and cruelty of a Genghis Khan, the sadistic mad genius of a Hitler, were, we hoped, temporary horrors in a slow development averaging toward the better. But while we remain limited to the surface of a gradually depleted Earth, we face a new kind of threat: even our success becomes failure. Survival will require that either voluntarily or under coercion we must limit our options. Heilbroner has argued that those limits will almost surely be more than physical, and that in the long run the freedom of the human mind will have to be limited also, as it is, very severely, in primitive human societies which have achieved stasis by a rigid social code.

We are surely far from having found the best ways in which human beings can live together and govern themselves; surely far from having achieved freedom for all, or having explored all the talents of which the human mind is capable. What chance will we have, though, here on an Earth ever more crowded and more hungry for energy and materials, to allow for diversity, for experiment, for groups to try in isolation to find better lifestyles? What chance for rare, talented individuals to create their own small worlds of home and family, as was so easy a century ago in our America as it expanded into a new frontier? For me, the age-old dreams of improvement, of change, of greater human freedom are the most poignant of all; and the most chilling prospect that I see for a planet-bound human race is that many of those dreams would be forever cut off for us.

·4·

NEW HABITATS FOR HUMANITY

Biologists and botanists talk of the "range" of a species: the limits, on the surface of Earth, over which a species can survive, grow, and reproduce. For our ancestors of the remote past, the range was the tropical ocean; it was a major step in the development of living beings when the early amphibians evolved into air-breathers. Now, when

41

we are about to design new habitats for man, we must question what limits are set by our own physiology. As we ask those questions, we must be conservative in our answers; we are not asking for extremes: not for the limits that apply to highly motivated athletes in superb physical condition, to mountain climbers, astronauts, or deep-sea divers, but for those that apply to quite ordinary people— ultimately, to "Aunt Minnie in her rocking chair." That conservative approach should apply even to the first habitat we build, for a practical reason that has a basis in hard economics: when people are called upon to work under hardship conditions, in miserable climates or ex- posed to disease, they have every reason to leave their families at home, and to demand high pay for their hardships and deprivations. Pay scales on the Alaskan pipeline construction job have to be very high. Even our first space colonies must pay their way, and they can only do so if they do not price themselves out of their markets. They must be places to which people come by choice, and to which their families enjoy coming also: places where it will be possible to live and work and raise children in ease and comfort.

With this conservative approach, we must then ask what constitutes a human environment; what is the "range" of mankind as a species? Most of us are accustomed to living near sea level. A large fraction of humanity, though, in mountainous areas of every continent, lives at altitudes as high as Denver, Colorado, where the pressure is 20 percent lower; and that fraction includes people who are elderly—a slightly lowered pressure doesn't seem to bother them.

The Federal Aviation Agency, to assure that pilots will be in a state of full alertness, requires that oxygen be used for any flight above 12,500 feet lasting more than a half hour. As a sailplane pilot, with my oxygen mask always at hand, I like to take a few breaths of oxygen at the tops of Western thermals, which are often a good deal higher than that. Serious mountaineers, climb- ing by muscle power and carrying packs, go far higher without oxygen, some to as much as 25,000 feet. Few human habitations, though, are more than twice as high as

Denver, and in those few, within the Andes and the Himalayas, the population has adapted through natural selection over many generations to life at low pressure. In space-habitat regions where people may be called on to do very light work not lasting more than a few hours, we can take the Federal Aviation Agency's limit as a guide, and for conservatism we should probably maintain in space-habitat living areas an oxygen pressure at least as rich as Denver's, a mile high.

As deep-sea divers and astronauts have shown, the nitrogen that makes up most of the atmosphere is unused by our bodies. On Earth, nitrogen serves to inhibit flames, and acts as part of our cosmic ray shield, but we do not consume it except through the food we eat. Curiously, neither do many plants: they take up nitrogen through their roots, from the soil, rather than from the air. If we provide some alternative way to inhibit flame and to protect ourselves from cosmic rays, the range of the human atmospheric environment will go as far as an oxygen atmosphere with the same oxygen pressure that is found in Denver. Though astronauts have lived in such atmospheres for several days while on the lunar surface, long-term tests with larger numbers of people will be needed before we can be sure that no respiratory problems will develop.

First we considered air, the medium without which we would be dead in a few minutes; next we can think of the range of temperature and climate over which humans can live and work. That range is wide, from the deep-freeze of the "Pole of Cold" in Siberia to the heat of the Sahara in midsummer. The range of comfort, and of easy operation without heavy clothing, is much narrower—just the few degrees where we set our room thermostats when we have the choice. Outside that range our efficiency goes down, and the steady migration to regions of mild climate without great variations suggests that the human desire for a comfortable temperature runs deep. We'd better plan on a narrow temperature range for most human activities, but allow for the variations needed for sports like skiing.

With atmosphere and mild climate, we can survive for

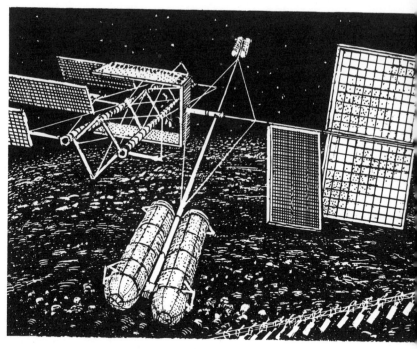

First normal-gravity habitat and
workshop in space, for pilot-plant
operations in low orbit.

one or two days; without water, though, we cannot last much longer. Nearly all the mass of our bodies is water, and in desert areas the inhabitants seldom deal with more than a few pounds of extra water per person. We are looking toward a pleasant, not a parched environment however, so will be much more generous: for the moment we will think in terms of several tons of water per person.

In extreme conditions people can go for several weeks without food, but in the space communities there will be no difficulty in providing food of a richer abundance, and with greater reliability, than exists over most of Earth. Water and food are no limits on the range of the human species in space.

Zero-gravity requires acclimatization, and for some people the adjustment takes several days. All three men of one Skylab crew were ill during the first twenty-four hours. Skylab tested a small sample of very healthy human beings for 90 days; during that time their bodies underwent definite physiological changes: a loss of blood volume, degeneration of certain bones, loss of bone marrow, and a slackening of muscle tone. Those changes were reversed and recovery was complete after some weeks when the men returned to Earth, but the advisability of exposing people to zero-gravity for many months without change seems doubtful; it's likely that a heart which had grown used to the easy conditions of zero-gravity might be prone to failure when gravity was restored. We do not want to make emigration into space a one-way trip, without the option of return at will.

Curiously, we all have the experience of what amounts to zero-gravity every day of our lives: the physiologists have found that bed rest takes the load off the body at least to the same extent as does zero-gravity, and that all the same types of degenerative changes occur in the two cases. We know that it is not necessary to be subjected to one gravity all the time; a few hours each day may be quite enough. How much less, we don't yet know, but it seems wise to plan that the areas where people will spend their time when they are not working will be at approximately Earth-normal gravity: ordinary people won't put up with the Skylab substitute for it, which was an intense program

46

of exercise occupying more than an hour every day. Fortunately again, gravity is easy to find in space: rotation can provide it. On the inside of a hollow, rotating vessel the gravity can be made to be the same as on Earth, and if the vessel is big enough the human body will find the artificial gravity indistinguishable from the real thing.

On Earth, sensitive, delicate organs within the inner ear have evolved to measure changes in the position of our bodies. Although they have their limitations, these organs can detect rotation about any of three axes.

Within a rotating environment, with a rotation period measured in fractions of a minute rather than twenty-four hours, our motion sensors can detect the fact that "all is not normal" as far as gravity is concerned. For a number of years, physiologists have conducted studies to find how difficult it will be for people to adjust to a rotating environment. The principal centers for these studies have been the U.S. Naval Medical Center at Pensacola, Florida, and the Soviet space program's ORBIT centrifuge facility in the U.S.S.R. Although there are limitations to the completeness with which such Earthbound tests truly duplicate conditions in space, there appears to be general agreement on the following points: first, almost no one has any difficulty in adjusting to rotation rates of one per minute or less. Second, as the rate climbs above two, three, four rotations per minute and even higher, more and more people find it difficult to adjust; they experience a variety of unpleasant symptoms ranging from motion sickness to drowsiness and depression. Some, though, are able to adapt to rotation rates as high as ten rotations per minute. In the case of a habitat in space, the range of interest is between one and three rotations per minute—high enough to be of concern, but low enough that most of the subjects so far tested have been able to adapt to it, usually within a day or two. For the larger habitats, which will almost surely follow the first small "models," the rotation rates can be kept below one rotation per minute without compromising efficiency of design. For the earliest habitats, economy appears to dictate that a rotation rate of about two RPM be chosen, for Earth-normal gravity, and that the applicants for jobs in the early habitats undergo tests to determine whether they are unusually vulnerable to motion sickness in space. The evidence from United States and Soviet space programs so far is that there is very little correlation between motion sickness as we encounter it in aircraft and boats; and the sort of "space sickness" that may be found when we substitute rotation for natural gravity. On the basis of the tests at Pensacola and in Russia, we can guess that only a few percent of the applicants for positions in the early habitats may find, after a few days or weeks in a low-orbital space station, that they are unsuited to life in space.

We have talked of the necessities of life, but if we are to work and live in space by choice, and enjoy doing so, we will ask for more: the age-old human desires of comfort, good food to eat and good wine to drink, room to stretch our legs, good places to swim and to get a suntan, variety in travel and amusement. We humans have definite ideas of our needs for enjoyment and amusement, and any successful space community will have to accommodate them.

We evolved as a hunting/gathering species, in the light of the sun, and our bodies need some exposure to it for well-being. Without sunshine, children develop rickets, and without sunshine people tend to grow moody and depressed: almost surely the high suicide rate characteristic of the Scandinavian nations is, at least in part, connected to cloudy skies and long cold winters. A successful space habitat will have to admit natural sunshine; that should not be hard to arrange; in space, remote from any planetary surface, full sunshine is available whenever we want it. But to avoid throwing off our internal biological clocks, evolved in a twenty-four-hour day, we will need to provide a day/night cycle.

When humans existed in small bands, they camped and always stayed near clear running water; except for their own smoky campfires the air they breathed was clean. In our pollution-ridden world no longer can we take clean air and water for granted; most large rivers are dirty. In a space habitat we should make a fresh start, and set up our industry and economy to keep the air and water clean.

Our Earth is rich in plants and animals, but as industry

and the human population crowd environments it is not as rich as it once was. City children become starved for the sight of a tree, and in desert areas the palms of the oases have an importance no dweller in a lush climate can imagine. For our psychological well-being, as well as for the cycling of the oxygen we breathe, we should have grass, trees, and flowers. Many animal species are a pleasure to us, and if we move into space both we and they will benefit by our taking them along—perhaps, like Noah's passenger list, two by two. Along with the domestic animals, we will certainly want to bring squirrels, deer, otter, and many others; birds, and some types of harmless insects for them to eat. In space, though, we have an option that does not exist to us on Earth: to take along those species which we want and which form parts of a complete ecological chain, but to leave behind some parasitic types: how delightful would be a summertime world of forests without mosquitoes! Perhaps, too, we can find less annoying scavengers than the housefly, and can take along the useful bees while leaving behind wasps and hornets.

Perhaps because we were originally a hunting and gathering species, the urge to travel and to seek out variety in habitat and environment is deeply rooted in many of us. Now that long-distance jet travel has become commonplace, a large segment of the population in the developed nations travels regularly for vacations. Our young people are learning wide horizons at a much younger age than did their parents. Some of the results are unattractive: traveling drifters, subsisting on doles from home and roaming the world as what the East-bloc nations call parasites; but if we believe in humanity we must also believe that the widening of horizons and the interaction of different lifestyles is, on the whole, a good thing: that it tends to cut away the hostilities and the myths that go with isolation, and so tends to reduce the likelihood of wars. Freedom to travel is precious, and adds greatly to human options; its blockage by poverty or by dictatorial governments always constitutes a loss. We can be grateful, then, that the technical imperatives of the humanization of space are toward easy travel at low cost; we cannot prevent the occasional abrogation of that freedom by a suspicious or reactionary government, but we can at least make sure that no barriers of poverty or energy-shortage act to prevent travel.

The growing of food is the most vital of all our industries, and now that we are freed of the planetary hangup we must ask: What are the optimum conditions for agriculture?

An adequate source of clean fresh water must always be at hand; in a space habitat, water once introduced can be recycled indefinitely, given an inexhaustible source of cheap energy.

The uncertainty of the Earthbound climate is the great bane of all farmers: drought, frost, or long-continued cloudy weather can ruin crops. Worse still, farming has always been subject to the cycle of boom-and-bust: in a good year, every farmer grows too much, and prices drop for his produce. In a bad year, he has little to sell although prices are high, and the consumer must pay highly for poor quality. In a space habitat, although people may want to live in climates that vary widely, crops should be grown in constant conditions, dependably unchanging from year to year.

Throughout most of the world only a part of the year is suitable for growing, and when winter strikes it stops all farming over a distance of thousands of miles. If we have a choice, we should provide that agricultural areas, in close proximity to each other and to the consumer, have the seasons and seasonal variations that are best for their particular crops. To make sure that our tables have fresh vegetables and fruit at all seasons, our growing areas should be staggered in phase—January in one while there is June in another. Impossible though that is on Earth, it will be easy in space.

On Earth, all of our high-yield grains, all of our fruits and vegetables, are subject to attack from various pests and viruses. Usually, these pests have evolved through centuries to attack certain plants, and on Earth winds and human travel threaten always to spread plant diseases to new areas. In space, it makes sense to start our agriculture with carefully inspected, pest-free seeds, and to introduce only those bacteria essential for plant growth. If our agricultural areas are separated from our living areas by even a few miles, and receive only sterile water and chemical fertilizers, the vacuum of space will serve as a perfect barrier to keep them pest-free: for the first time, we will be able to have agriculture of high yield without pesticides, insecticides, or crop losses due to raiding birds and animals.

As agriculture has become more and more sophisticated, it has become ever more factorylike. In modern high-yield agriculture, the soil in which crops are grown is relatively unimportant; it serves only as a matrix to hold the growing plants. The highest yields are obtained by intense application of chemical fertilizers, and by careful control of trace elements and the acidity of the soil. As the evolution from a pastoral economy to an agricultural industry has gone on, that industry has become continually more energy-intensive. The cost of fertilizer production is dominated by the cost of energy.[1] In space a method for the production of fertilizer will become easy, though on Earth it is uneconomical: that is the simple heating of an oxygen-nitrogen mixture, in a tube at the focus of an aluminum-foil mirror in sunlight, to a white-hot temperature. At that heat about 2 percent of the molecules will dissociate and recombine to form nitric oxide, an energy-rich precursor of chemical fertilizer.

It appears, therefore, that space can provide the ideal conditions for a highly efficient, totally recycling agriculture no longer at the mercy of weather and climate.

We are examining the needs of an industrial civilization, so we must look toward the conditions in which industry can work efficiently, at low cost, free of pollution.

Industry is energy-intensive, and with increasing sophistication and the continuation of the industrial revolution, that hunger for energy also grows. Here on Earth, where our energy sources are limited, we have come to think of intense energy-usage as very nearly immoral; but if we have a truly unlimited source, there is no reason to curtail the natural development of the industrial revolution.

Industry uses energy in two forms: electrical and ther-

Construction of Island One,
seen past completed farm area.

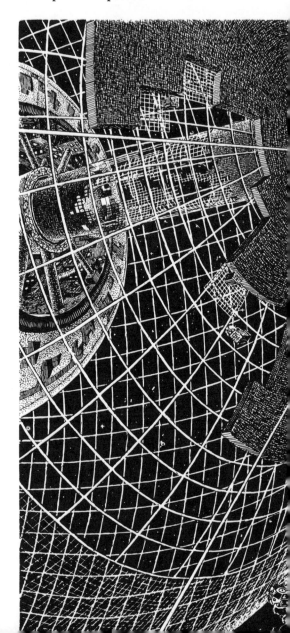

mal. Thermal energy is used for melting metals, for raising chemicals to temperatures at which they react, and for making ceramics. On Earth most of the fossil fuels that industry uses are burned to provide this thermal energy. In zero-gravity, far from a planet, the concentration of the unvarying, intense sunlight of space by very lightweight, inexpensive mirrors can provide all the energy that industry will ever need. A simple reflector the size of a football field, weighing no more than a car, when extended in space can provide a great deal of process-heating; to equal it an Earthbound factory would have to burn a million barrels of oil every thirty years—but the reflector in space will go on supplying that same power at no cost, as long as the Sun shines.[2]

I spoke of the ease of obtaining electric power in large quantities from sunlight in space. We can be more quantitative about it: given an industry in space, at which large turbogenerator power plants can be built, we can expect to build them for about the price of a coal-fired plant on Earth.

The space power plant, running at zero-gravity, will need less maintenance than its Earthbound counterpart; even though its turbine rotor and generator armature may have a mass of thousands of tons, they will weigh nothing in zero-g, and can be supported with no direct frictional contact, on air or magnetic bearings which should have an infinite lifetime. The fuel cost for a plant in space will be zero, so the entire cost of power will be that of amortization, maintenance, and distribution. In space the industries that use electric power can locate anywhere in a volume, rather than on a flat surface, so they can be much closer to the power plant, reducing distribution costs. Maintenance should be low, because there will be no fuel-handling machinery to service and no friction bearings to wear out.

Putting all the numbers together, a turbogenerator plant running on solar energy in space should be able to supply electricity to nearby industries at a fraction of a cent per kilowatt-hour; that figure is lower than the cost of electricity in all parts of the U.S. except where hydroelectric power is available. After amortization, costs should drop to those of maintenance. The cost of power enters into every part of an industrial economy, so in space it should be possible to produce most goods more cheaply than on Earth.

There is an additional component to energy costs, a component whose force we are starting to appreciate: the cost of uncertainty. When the planners of a new industry cannot predict how much electric and thermal power is going to cost at the time a new facility will be finished, they find it very difficult to make the decision to build, and even more difficult to persuade a lending agency to advance the money for construction. In space, that uncertainty will be removed, because fuel costs will be zero and can be guaranteed to remain so for the life of the Sun: several billion years at the best estimate. Lloyds of London should be very willing to insure a new industry against its power costs going up, with that kind of backing!

We should examine whether nuclear fission or fusion power on Earth can ever equal the low costs of solar power in a space colony. The answer seems to be, No: Earthbound nuclear power will not compete successfully with solar power in space. First, for all process-heating needs, in space a simple mirror with no moving parts, located at the point of use, will be sufficient. On Earth one would have to go through the expensive and inefficient intermediate step of converting from nuclear power to electricity and then back to heat, since nuclear plants cannot be made in small sizes. For electric power on Earth from fusion, we will overlook for a moment the fact that billions of dollars and twenty years of effort have so far failed to make nuclear fusion a practical reality. Even if it succeeds, its cost will almost surely be much higher than those of a solar plant in space: in a fusion plant, one will first have to spend energy to separate the one part in 5,000 of heavy water from ordinary water, then obtain deuterium from it. Then it will be necessary to pass through a stage of complicated, high-technology machinery, involving either lasers or giant magnets. In the end, one will have heat—only to put it into the boiler of a turbogenerator plant. The space-borne solar plant will bypass all the hard part of this complicated sequence, because it will begin with free solar energy.

Finally, the distribution costs in space will be far lower, because distances from power plant to industry will be only a few miles, and because solar electric plants, unlike nuclear stations, can be made in small, convenient sizes adjacent to heavy power users.

In addition to the advantages of zero-gravity for the handling of massive objects, for the heating of materials to high temperatures without the contamination of confining crucible-walls, for the formation of uniform mixtures of heavy and light materials,[3] and for the growing of large single crystals, industry in space will have an additional degree of freedom. By gentle rotation, it will be possible to maintain very thin surfaces accurately in the form of cylinders and cones. That may be especially useful in the case of large mirrors made of thin foil.

Here on Earth our lowest-cost transportation is that of crude oil in supertankers. Though the rates fluctuate wildly, tanker construction being about as speculative as pulling the handle on a Las Vegas slot machine, the bare operating costs amount to about 0.06 cents per ton-mile.[4] For shipment of commodities in bulk from one space-colony to another, at a speed typical of highway driving on Earth, a tanker-size payload can simply be put in one large motorless container, and accelerated by an electric motor and cable to its drift speed. No crew need go with it, because in the vacuum of space its trajectory and its time of arrival will be known exactly, and there will be no weather or navigational hazards to contend with. The energy cost of such a shipment will be absurdly small: only about a thousandth of the cost per ton-mile that a supertanker works for on Earth.

Commuting to work from a space-colony should be correspondingly easy and inexpensive. The typical vehicle can be a sphere, protected from cosmic-rays by a dense, foot-thick outer shell. It may contain seating on three levels, and be entered by three airliner-type doors. With a comfortably generous amount of elbowroom and legroom for each passenger, about like those of first-class seating on a long-distance airliner, the sphere can accommodate a hundred passengers. In less than a half-minute, an electric motor and cable can accelerate the sphere to the speed of a

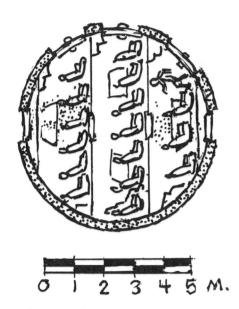

Commutersphere used on
short-distance daily trips.

jet plane, and the flight to a factory a hundred miles or so
from the colony will take only a few minutes of vibration-
less flight. Just time enough to skim the morning news,
and an arresting cable will slow the sphere to its destination. The energy cost? Less than fifty cents per passenger.

Each time the balance is tipped for a particular industry,
so that production in space becomes cheaper than on
Earth, we will be relieving Earth in two ways: we will be
removing the burden of energy usage and materials mining for that industry, and we will be generating an additional force to draw away population: the work force of
that industry, and the families of the work force. For many
years, the only industries in space that will compete
directly with those on Earth will be industries that require
no material shipment of material products back to Earth;
there are at least two of these: fabrication shops to produce
satellite solar-power stations, for location in geosynchronous orbit above a fixed point on Earth's surface to beam
down power for Earth's electric systems; and assembly

plants for the aerospace industry, building ships for transport among the colonies and from Earth out to the colonies.

For energy in the United States alone, we now burn literally billions of tons of irreplaceable fossil fuels every year. From a conservation viewpoint, it makes little sense to blow away this oil and coal in the form of smoke; it should probably be conserved for use in making plastics and fabrics. That environmental consideration, reinforced by a powerful economic drive, suggests the construction of solar-power stations for Earth as perhaps the first major industry for the space colonies.

Within the colonies themselves, no conflict need ever arise between using carbonaceous materials for energy and using them as they should be used: for the petrochemical industry. As we have seen, the cost of solar power in a space colony will be so low that it will be ridiculous there to obtain energy in any other way.

For the continued growth of wealth, a developing economy must have an assured source of materials. On Earth, we are already forced to work poorer sources to obtain our metals; for iron in the United States we have long since depleted the Mesabi range in upper Michigan. As we work poorer veins, the conflict of mining with the environment rapidly becomes more serious: when the ore content is only a tenth of that in a rich vein, we must mine and process ten times as much material to get the same quantity of the metal we seek.

In space, our first mines will almost surely be on the Moon. Particularly on the lunar Farside, enormous quantities of materials could be removed without ill effects of any kind. It comes as a surprise to most people to learn how rich a source of industrial materials the Moon is; I believe that in the long run the Apollo Project, much criticized as it was during its lifetime, will be seen to have been of enormous value for its lunar prospecting function. A typical Apollo sample contains, by weight, more than 20 percent silicon, more than 12 percent aluminum, 4 percent iron, and 3 percent magnesium. Many of the Apollo samples contained more than 6 percent titanium by weight; titanium is in great demand as a strong, light metal, which holds its strength up to a very high temperature. Its present use is mainly in the aerospace industry; processing it requires high vacuum, high temperature, and a lot of energy: all things which are expensive on Earth but will be cheap in space. Finally, the lunar surface is more than 40 percent oxygen by weight; strange to think that such a lifeless, sterile landscape contains, locked in its soils and rocks and waiting to be used, the one element we need most to sustain our lives.

In the "long" run, within one or two decades after the human use of space begins, we will begin to exploit the resources of the asteroid belt. For transport in space we must think in terms of energy rather than of distance, because travel in space is without atmospheric drag. To bring a ton of material, efficiently, from Earth's surface to the site of a space community would cost about the same, in energy, as to bring that ton of material to the same point from the asteroid belt. The difference is that lifting it from the surface of Earth requires a rocket able to supply more than a ton of thrust, and further requires elaborate fast-acting control systems operating with split-second precision. By contrast, moving a load of freight from the asteroids to the colonies can be done in a leisurely fashion, with efficient, low-thrust engines. If the engines break down, there will always be plenty of time to fix them, just as a freighter on Earth's oceans can lie dead in the water for days if need be while its engines are being repaired.

Bringing materials from the lunar surface to the site of the space communities will be even easier; the energy cost per ton will be only about one twentieth as much as for shipment from Earth or from the asteroids. As we shall see in later chapters, materials can be brought from the Moon at an initial cost of only a few dollars per kilogram. Later, when space-borne industry is well established, the ultimate costs should drop to only a few cents per kilogram.

The Moon is poor in three elements which we need for life and for a full industrial base: hydrogen, nitrogen, and carbon. Apparently during its lifetime the Moon was subject to baking at a very high temperature. Fortunately, it has been shown by analyzing the spectra of sunlight reflected from asteroids that some of them are rich in carbon, nitrogen, and hydrogen; they are about as good a source for petrochemicals as oil shale.[5] Corroborating evidence for the presence of these elements in the asteroids comes from about twenty meteoroids found on Earth's surface;[6] they are of a type called "carbonaceous chondritic." The normal economic decisions that govern industrial operations will therefore probably lead to mining the lunar surface for most elements, and the asteroids only for the materials which the Moon lacks. Long before an appreciable fraction of the lunar surface has been mined, it will become easiest to obtain all the materials for colony construction at the asteroids themselves.

Although the total volume of the asteroids is far smaller than Earth's, it is a volume much more accessible than the depths of our planet. On Earth only a thin skin of material is available to us without deep mining under high pressures and intense heat. Even if we were to excavate the entire land area of Earth to a depth of a half-mile, and to honeycomb the terrain to remove a tenth of all its total volume, we would obtain only 1 percent of the materials contained in just the three largest asteroids. A striking contrast: we would have to disfigure the entire Earth to obtain only a hundredth of the material contained in three now-useless, lifeless asteroids; and there are thousands of those minor planets. Moreover, to bring material into space even from the biggest asteroids requires climbing a gravitational hill only five to ten miles high, instead of Earth's 4,000 miles.

As a reader of science fiction in childhood, I gained no clue that the future of mankind lay in open space rather than on a planetary surface. Later, when logic and calculation forced me to that conclusion, I searched for evidence that others before me had come to the same realization. More than five years after my studies on this topic began, I found the references I needed: a friend obtained for me copies of two books, out of print in their English editions, by the self-educated Russian scholar Konstantin Tsiolkowsky.[7,8] Born in 1857, Tsiolkowsky wrote pioneering works on reaction motors, multistage rockets, and many other basic concepts of the space age.

60

Tsiolkowsky's novel *Beyond the Planet Earth*, written at the turn of the century, serialized, and finally published in book form in 1920, is a thinly veiled treatise on basic physics. As such it is short on characterization, and should be read for what it is: a daring but logical feat of the imagination. At a time when transportation was still almost exclusively horse-drawn, it required a bold thinker indeed to speak casually (and accurately) of the necessary orbital speeds of kilometers per second.

As a novelist, Tsiolkowsky could skip lightly over the problems whose solution he could not then see: the rocket on which his voyagers lift off from Earth is powered by a mysterious explosive of a nature left unexplained; but the circumstances of the flight show surprising parallels to our present predicament on Earth. Tsiolkowsky postulates an Earth on which a growing population is beginning to feel the ecological limits. His travelers visit the Moon only incidentally; they realize from the start that the place for settlement is well away from any planetary surface:

"Meanwhile the new colonies, five and a half Earth radii or 34,000 kilometers away, grew and were peopled. Mansion-conservatories of the type we have described were filling up with fortunate men, women and children. . . ."

They see the advantages of free space for establishing gravity convenient for particular tasks:

". . . nothing could be simpler than to create it artificially, you see, by rotating the house. In space, once you start a body rotating, it goes on rotating indefinitely, there is no effort involved; so the gravity is also maintained indefinitely, it costs nothing. Moreover, the amount of gravity depends on us; you can make it lower than terrestrial gravity, or higher."

On their first flight, Tsiolkowsky's travelers foresee accurately many of the possibilities of industry and habitation in space:

"The space around the Earth which we can use—assuming we count only half the distance to the Moon—gets a thousand times more solar energy than the Earth . . . it only remains to fill it with dwellings, greenhouses—and people. By means of parabolic mirrors we can produce a

61

temperature of up to 5,000 degrees centigrade, while the absence of gravity makes it possible to construct mirrors of virtually unlimited size, and consequently to obtain foci of any area we choose. The high temperature, the chemical and thermal energy of the Sun's rays, not weakened by the atmosphere, makes it possible to carry out all kinds of factory work, such as metal welding, recovering metals from ores, forging, casting, rolling, and so forth."

Sensibly enough, the travelers spend much of their first voyage in a search for usable asteroids. As a novelist, Tsiolkowsky has no difficulty in filling the asteroids with gold, platinum, and diamonds, but in our more practical day we will be glad enough to find there such homely elements as carbon and hydrogen. Of all the prophecies Tsiolkowsky made during his long life, I am glad that one in particular was selected for the obelisk marking his grave in Kaluga:

"Man will not always stay on Earth; the pursuit of light and space will lead him to penetrate the bounds of the atmosphere, timidly at first, but in the end to conquer the whole of solar space."

5
ISLANDS IN SPACE

While we are considering the form of the new habitats for humans—the islands in space—we must always remember that details will change, perhaps profoundly, between earliest conception and final realization.[1,2] There may be better solutions to some technical problems than those I outline, and also new problems may arise whose solution will require changes in the design. I am describing a kind of "existence proof"—an illustration that a consistent solution does exist for the design of the islands in space; but it would be strange indeed if the efforts of one man were not improved upon greatly when others consider the problem.

I confess to a humanitarian bias in the design that I suggest. Technological revolution is a powerful force for social change, and in choosing among several technical possibilities I have been biased strongly toward those which seem to offer the greatest possibilities for enlarging human options, and for breaking through repressions which might otherwise be unbreakable. Yet I offer no Utopia; man changes only on a time scale of millenia, and he has always within him the capacity for evil as well as for good. Material well-being and freedom of choice do not guarantee happiness, and for some people choice can be threatening, even frightening. Though I acknowledge that my study will be of the physical environment, and only indirectly with the psychological, I will still try to describe an environment which combines with its efficiencies and its practicality opportunities for increasing the options, the pleasures, and the freedoms of individual human beings.

I have argued that there is only one way in which we can develop truly high-growth-rate industry, able to continue the course of its development for a very long time without environmental damage: to combine unlimited solar power, the virtually unlimited resources of the Moon and the asteroid belt, and locations near Earth but not on a planetary surface.

I will describe first a community of what I like to call "moderate" size; it is larger than the first model habitat, but far below the dimensions of the largest that might be built. "Island Three" is efficient enough in the use of materials that it could be built in the early years of the next century. The numbers will seem staggering, but they are backed by calculation: within the limits of present technology "Island Three" could have a diameter of four miles, a length of twenty miles, and a total land area of five hundred square miles, supporting a population of several million people. The largest communities that could be built, within the limits of ordinary, present-day structural materials like iron and aluminum, and with oxygen pressures equal to 5,000 feet above sea level on Earth, could be as much as four times as long and wide, with a land area half the size of Switzerland. It would be uneconomic at first to build habitats that large; they would be wasteful of materials. In the long run, though, the human race may build habitats of that size, or, with more advanced technologies, even larger.

We need to provide gravity, water, land, air, and natural sunshine in an Earthlike environment. Rotation can simulate gravity, and fortunately there are at least two geometries that allow rotation while giving us the real Sun stationary in the sky. One is a coupled pair of cylinders, whose long axes are parallel to each other. The cylinders are closed by hemispherical endcaps, and contain oxygen. Each cylinder rotates about its long axis, so that people living on its inner surface feel an Earth-normal gravity.

The cylinder circumference is divided into six regions, three "valleys" alternating with three arrays of windows. By locating three large, light planar mirrors above the windows, and pointing the cylinder axes always toward the Sun, we can arrange that the valleys will receive natural sunshine, and that the Sun will appear motionless in the sky even though the cylinder is rotating. Varying the mirror angle will give dawn, the slow passage of the Sun across the sky during the day, and sunset. The day-length, weather, seasonal cycle and heat balance of the colony can be regulated by the same schedule of mirror-angle variation. A large paraboloidal mirror at the end of each cylinder can be collecting solar energy twenty-four hours per day, to run the community's power plant.

If we then set up many smaller cylinders near the big ones, and use the small ones for the growing of crops, we will achieve what has never been possible on Earth: independent control of the best climates for living, for agriculture and for industry all within a few miles of each other.

The "valley" areas, in Island Three, would each be two

65

Island Three, with agricultural modules and zero-gravity industries.

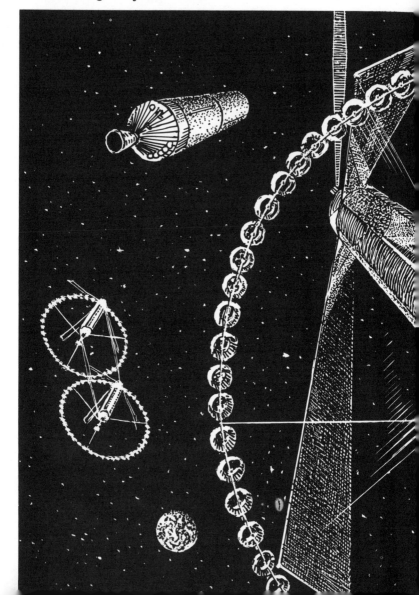

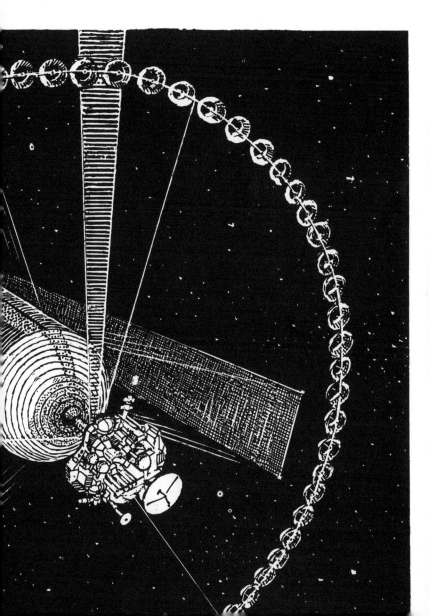

miles wide and twenty miles long, rising beyond that to mountains. These mountains, formed on the inner surfaces of the cylinder endcaps, could have a height of up to 10,000 feet.

In the simplest version of a space-community design, sunlight will be reflected into the habitat by large plane-surface mirrors, attached by many cables to each rotating cylinder and rotating with it. A dweller in one of the valley areas will look up and see a blue sky, obtained probably by art rather than by nature: it will be rather easy to control the reflectance of the mirrors and the tinting of the windows areas ("solars") to produce the most pleasing combination of warmth and brightness for the sunshine falling on the valleys, and to give a blue tint to the solars. There will be no sensation of rotation, though the cylinder will be turning once every two minutes; gravity in the valley areas will be Earth-normal. No one in the space habitat will be in any doubt as to where he is, though: high above him, far above the clouds he will see, dimmed by distance, the other two valleys of his home. From that far away they will be as indistinct in detail as the Earth's surface is from an aircraft four miles high, but the inhabitants will be able to see them.

The angle of the sunlight entering the habitat will be controllable, and will depend only on the lengths of the cables which hold the mirrors. As the mirrors slowly open in the morning, the Sun will rise, but will move in the sky only as fast as it does on Earth; there will be no suggestion from its appearance that the cylinder is actually rotating. Only with very delicate instruments could one find that the image of the Sun's disc is rotating around its center.

With control over the angle of the Sun in the sky, the residents of space will also have control over the lengths of their days, the variation of the day-length, and so the average climate and the seasons. They are unlikely to indulge in any sudden or capricious changes in those variables. Humans can adjust quickly, as the jet age has shown us, to changes in the day/night cycle and the climate; plants and trees, though, are not so adaptable, and once a cycle has been established there will be good reason to make changes in it only very slowly.

By the time a community as large as Island Three is built, space habitats may not be occupied at the ecological limit: the highest population density that the land can support. In the early years of the next century Earth will be from two to three times as crowded as it is now, and the population density in space habitats may be falling toward the same value as that of Earth, ultimately to cross it and fall still lower. Island Three however could support quite easily a population of ten million people, growing its food in agricultural cylinders near but outside the main habitat. In the figure of habitat cost per person, I will assume that higher density. We are used to the perpetual conflict, here on Earth, between industry, agriculture, and living space, but we must realize that in a space habitat economics will dictate escaping that conflict by locating agriculture a few miles away from the living areas. It is relatively expensive in materials to build large cylinders, with diameters of several miles, and relatively expensive to provide sunlight of normal appearance. Plants do not need such luxuries, and can be grown very efficiently in places where the solar intensity is high, but where there are no visual amenities.

With industry and agriculture located outside, the dwellers in Island Three can use their two hundred and fifty square miles of land area for living space and recreation. I suspect that as colonists from various countries of Earth arrive to settle the many communities in space, there will be a great variety in the ways in which land area will be used. Some immigrants may choose to arrange their land area in small villages, with single-family homes, the villages being separated by forests. Others may prefer to build small, intimate towns of high population density, to enjoy for example the color and excitement and human interaction that is so much a feature of small villages in Italy. With many new communities to choose from, the emigrants from Earth will settle in those they like best. I would have a preference, I think, for one rather appealing arrangement: to leave the valleys free for small villages, forests, and parks, to have lakes in the valley ends, at the foot of the mountains, and to have small cities rising into the foothills from the lakeshores. Even at the high-population density that might characterize an early

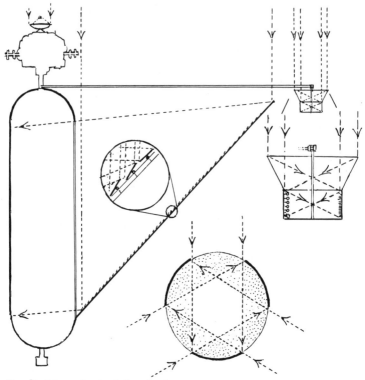

Sunlight paths in Island Three and
in its external farm modules.

habitat, that arrangement would seem rather pleasant: a
house in a small village where life could be relaxed and
children could be raised with room to play; and just five or
ten miles away, a small city, with a population somewhat
smaller than San Francisco's, to which one could go for
theaters, museums, and concerts.

For Island Three, taken as a community that might be
built in the first half of the next century, I assume that the
population density will be "high," though as I have
described it the habitat need not seem crowded. To feed
that population, it will be quite enough to have an agricultural growing area equal to that of the living habitat. That
may seem surprisingly small; it corresponds to growing

the food to support one person on a plot only about thirty feet on a side. On Earth agriculture never reaches so high a productivity. The number is based, though, on yields which have already been achieved in a remarkable series of experiments by a unique individual.

After a long and active career at Cornell University, Dr. Richard Bradfield retired in 1965. Soon afterward he came out of retirement and assumed a responsibility and a physical challenge that many younger men would have found too demanding: the directorship, under the sponsorship of the Rockefeller Foundation, of the International Agricultural Experimental Station in the Philippines. That Institute, an outdoor laboratory for the development of new methods in high-yield agriculture, is a world center for what has come to be called the "green revolution." Dr. Bradfield found that yields could be raised greatly by two expedients: multiple cropping and double planting.[3] In multiple cropping, advantage is taken of the fact that a high-growing crop, like corn, can thrive in the same rows in which a low crop, like sweet potatoes, is grown. As long as nourishment is supplied to both by intense application of fertilizer, the two crops can live and grow together in harmony.

Double planting takes advantage of a fact that even amateur gardeners know: in the first weeks after seeds are planted, their growth does not depend on sunlight, or even on nutrients; all they require is warmth and moisture. The technique of double planting consists simply of overlapping one growing cycle with another: for a fast-growing crop like hybrid corn, reaching maturity in just ninety to one hundred days, to plant the seeds of the next crop ten or twenty days before the last one is harvested.

By these methods Dr. Bradfield was able to reach very high yields even with what was basically conventional agriculture—not hydroponics. His agricultural station could support about twenty-five people per acre, even in the less than ideal Philippine climate.[3,4]

Using Dr. Bradfield's data, one can work out the yield of agriculture products for a space-community agricultural area, where the temperature will be ideal for growing (probably about like that of a hot summer day in Iowa) and the climate will be unvarying. Under such conditions there is no reason why four crops per year cannot be obtained.

In the developed countries we are used to a varied and probably too rich diet, but in the space habitats there is no reason why anyone should be limited to a diet of rice or cereals. Specialists in diet tell us how many calories and how many grams of usable protein we need every day if we're doing active physical work, and the space-agriculture areas are figured on that basis.[5] Many of us find that on such a diet we have trouble to keep from gaining weight. In the early space communities it will not be very practical to raise many beef cattle; they are quite inefficient at converting plant foods to protein-rich meat, losing a large factor in the process. Chicken and turkeys, though, are quite efficient, and pigs are not much less so. Dr. Bradfield has found that in high-yield agriculture the cuttings from crops like corn and sweet potatoes can be used efficiently as pig-fodder,[6] and that practice can be employed to good effect in a space community's agricultural cylinders.

With a varied diet including all the corn, cereals, breads, and pastries that many of us enjoy, and with plenty of poultry and pork, the space colonists will have good reason to follow our Pilgrim ancestors, and celebrate Thanksgiving with a feast of turkey, and Christmas with a savory ham. There will be no need for anyone to think in terms of pressed soybean cakes or fishmeal unless they happen to like such things. With four crops per year, a completely dependable climate free of hurricanes and frosts, and techniques that Dr. Bradfield has developed, the space communities can easily support at least fifty-three people per acre of agricultural land.

The agricultural areas of a space habitat will probably be relatively small, perhaps a square mile in area. They may be cylinders, but will not have external rotating mirrors; simple conical reflectors will be quite sufficient for them, since a stalk of corn will hardly care whether the image of the Sun is round or elliptical. The agricultural areas will probably be run at rather a low density of oxygen, corresponding perhaps to a mountain altitude, because that will make the enclosing structure cheaper and because plants grow best with less air. The climate will be hot and moist for most crops, and the day-length will be controlled in an inexpensive way, by drawing an aluminum-foil shade in zero-gravity, outside the cylinder, across the front of the mirrors to close off the sunlight. Looking into such an agricultural cylinder, one would rarely see a farmer, just as it is rare to see a human being as one drives through the San Joaquin Valley, one of our highest-yield agricultural areas in the United States. As in that valley, the occasional farmer will almost certainly be driving a machine of some kind: a planter or harvester; at the space community the machine may be air-conditioned, perhaps even pressurized, and shielded against the radiation from solar flares.

When winter comes to the San Joaquin, it closes down the growing season over the entire valley at once. That will not happen in the space habitat. There, each cylinder can have the climate and season that people choose, because those factors will be controlled by the moisture content of the air and ground, and by the schedule of day-length for each cylinder. With such control, there is every reason not only to tailor the climate of each cylinder to favor its particular crop, but to distribute the seasons among the cylinders. They may even run in serial order: January, February, March, and so on. With that degree of freedom it will be possible to have every desired crop always "in season" in one of the areas, so that the settlers living a few miles away can enjoy, for example, fresh strawberries even in the middle of what may be their January.

In the long run, when plentiful supplies of water are available from the asteroids, it will be possible to specialize certain of the agricultural areas as ponds and lakes, both fresh and salt. Oysters, clams, fish of all kinds, and perhaps even that vanishing delicacy the lobster, may all be grown there.

The practicality of these options depends on the free solar energy continuously available in space, and on the fact that such energy can be used for the inexpensive production of chemical fertilizer. A plant using direct thermal energy and converting oxygen and nitrogen to nitric oxide,

Interior of
Island Three;
view from hillside
toward valley.

adequate to provide abundant fertilizer for a space habitat, need have a concentrating mirror area of only a square meter per person, because of the high intensity of sunlight in space and its availability day and night, all year.

I like to be on the safe side in my estimates, and by the time space colonies become a reality it's likely that we'll be able to do even better than I've "promised." Detailed studies supported by NASA in 1975 and 1977, with participation by experts in high-yield agriculture, concluded that the numbers I've given are well on the conservative side. The General Electric Company feels sure enough about closed-environment greenhouse agriculture that in 1977 it committed corporate funds for a half-acre pilot plant, and is betting on yields even much higher than I've figured.

The space habitats will operate, of course, in a totally recycling way: fresh produce, fruit, vegetables, meat, milk, and cheese will travel the short distance from the agricultural areas to the living habitat, and the return flow will be pure water and nutrients for the fertilizer plants; nothing will be thrown away. Passing all wastes through a high-temperature solar furnace, which will cost almost nothing, will ensure that everything entering the agricultural areas is sterile. In that way they can be kept free of pests, even if any should accidentally be introduced into the living habitat. In the very worst case—the introduction or evolution of an agricultural disease in one of the areas—the sterilization process that will be part of the recycling will ensure that the disease will not spread. As soon as such a disease is found, there will be a simple and preferable alternative to the Earthbound necessity of sprays and poisons: it will only be necessary to drain off the water of the contaminated cylinder through a solar steam boiler to a sterile tank, and to open the shades so that the cylinder heats up to a temperature at which no living organism can survive. After a few days or weeks of that treatment the water can be re-introduced, the appropriate soil bacteria can be replaced, and a new planting cycle can begin.

The population density in the space habitats will be governed by sheer economics: there will be a certain cost per square mile of land area, low for the minimal, early

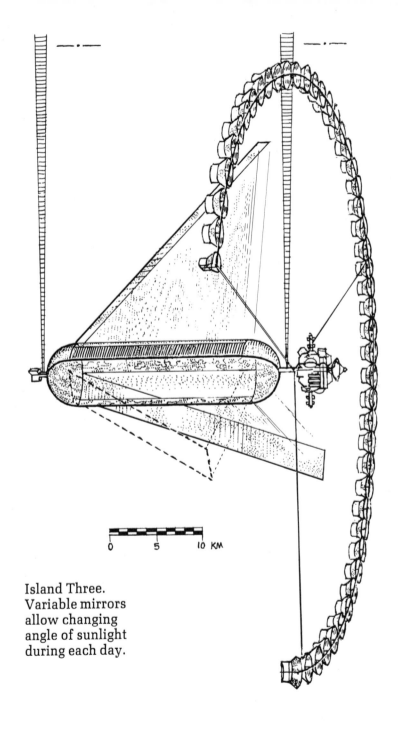

Island Three.
Variable mirrors
allow changing
angle of sunlight
during each day.

0   5   10 KM

communities, higher for the larger ones. A habitat of large diameter will require a thicker supporting shell of aluminum or steel. As I have emphasized, a key element in the humanization of space will be the unchecked continuation of the industrial revolution, the process by which average individual productivity and wealth increases. That increase translates into time, when we consider population density in a community: in the early stages it will not be possible economically to build and amortize a community unless it contains a large work force to pay off the construction costs within the amortization period. Later, as automation, productivity, and with it average wealth increase, it will become possible to build relatively large communities for habitation by comparatively few people. As we shall see in a later chapter, that transition will not take long; with a normal growth in productivity, less than a century will suffice for a reduction of a factor ten in population density.

Island Three, though, is taken to be an early model, built when productivity is still not very much greater than in a developed nation on Earth at the present time. It may have ten million people, and we should look at what that large population will mean in terms of living conditions.

With half the total population living in small cities on the mountain slopes, and agriculture carried out in the two hundred seventy square miles of external cylinders, the habitat valleys may be used entirely for green areas and for suburban towns. Though the life-styles may be as varied as the national origins of the colonists, one possibility is to have a series of small villages nested in a forest. A population of 25,000 in a village will be enough to support schools and shops, and such a town need be little more than a mile across. With predictably good weather and a mild seasonal variation, bicycles and small electric runabouts will be quite adequate for travel within the village, so it can be a place free of automobiles and of the internal combustion engine. Though I speak of Island Three as a high-density community, the living conditions in the village will hardly seem crowded: a family of five could easily have a one-story home of four- or five-bedroom size, with large living areas. They could have in addition a garden and a yard of equal area, while still leaving much of the village free for shops, schools, and perhaps a village green.

Some features of the habitat geometry will lead to new possibilities in house design. For one, the ubiquitous, ugly TV antenna of American suburbia will vanish, to be replaced by a small built-in concealed equivalent, pointing at the center of the cylinder endcap. With direct line-of-sight and a distance of only a few miles, reception should be superb. Probably by the time such a community is built families will also be able to communicate with a central library through the same microwave link.

Electric power, brought underground from the external solar-power station over cables laid in when the community is built, will run lights, appliances, and air conditioning. Most of the energy we use, though, goes into heat, for house heating and for cooking. In Island Three, all such directly used heat may be obtained from solar power,

Sunflower habitat, this one designed for Southwestern Desert landscape. A transitional geometry leading to Island One.

without ever going through an intermediate stage as electricity. The ground on which houses are built may be no more than two feet thick, and at the time of construction several access channels down to the outer shell may also be built in. Even at nighttime solar power will always be available, no more than a few feet from the house floor. Reflected by external mirrors, solar heat for cooking can be brought up through the floor through a window and through a short channel, to be absorbed on the lower side of a simple metal cooking surface. A rather powerful electric range element can be replaced by a cooking surface fed by just two square yards of solar collecting-mirror, and turned off by a simple shutter. The heating of each room in the home can be done in the same way. Whether that is the course adopted, or whether electric power will be so cheap that it will be used for all energy needs, is a matter for economics and design ingenuity to determine.

The homes of Island Three may also have a design detail no Earthbound home can ever equal: a window set at an angle in one wall of a living room, through which the immensity of space and the brilliant, unclouded stars will always be visible, drifting majestically across the field of view as Island Three rotates in its unvarying two-minute cycle.

The production facilities of Island Three may be of two kinds: light industry, located in the cities or even in the villages, and heavy industry, outside the habitat entirely. On Earth, industry must compete with us for land area on which to locate. But no such conflict will arise at Island Three.

An industrial complex located just outside an end of the community, and nonrotating, will be an ideal facility for processing lunar materials into finished products. At each end of each cylindrical habitat, there can be a thin, nonrotating disc of zero-gravity industries; a disc as big as the colony's end-cap, but only as thick as one factory. In that arrangement, each industry can have its own direct access to space, to receive raw-materials shipments and to send off its finished products. In such a geometry waste heat from these industries can be radiated away into the cold of outer space with equal ease. Workers in these zero-gravity industries can travel from the cylinder axis to their jobs in a few minutes, in a large air-filled zero-gravity corridor, pushing off from their starting points and drifting in free flight to their destinations—and possibly reading their magazines as they go.

The products of these zero-gravity industries could be very large indeed. There is no reason why an external factory couldn't build and fully assemble a complete solar-power station, which could then be gently floated away in zero-gravity to its destined point of use.

In the energy-rich environment of a space community, it will normally be more efficient to recover and separate industrial waste products for their useful materials, but if any smoke or gases do escape from a factory they will be carried by the solar wind all the way out of our solar system, never to add pollution to the environment.

In most of the agricultural areas of Island Three, except for the insects essential to pollination, there will be no reason to have animal life; no birds, for example, because they would attack the crops.

In the main living areas, though, we may find the ideal habitat in which species endangered on Earth may survive. There will be no need to introduce insecticides or other poisons, and industrial wastes, if any, will be borne away by the solar wind, never to enter the habitat itself. In those conditions, with choice in the species which are intro- duced to form the initial ecosystem, it may be quite possible to bring rare species of birds and animals from Earth to the nonagricultural areas, and to have them survive and flourish.

Every step toward the settlement of space will benefit conservation programs in another way: by relieving Earth of industry and of its burden of population, so that the species of animals, birds and fish now in danger on Earth will have a better chance of survival here.

6
NEW EARTH

A few hours of time with pencil and paper, while letting the imagination roam, will be enough to convince any reader that many geometries are possible for habitats in space. In the long run, it seems likely that in designing their environments dwellers in space will take full advantage of new degrees of freedom in gravity, day-length, and

Village in Island One.

climate. My reason for describing a much more conservative, narrowly Earthlike habitat is that all of us now on Earth, who must decide our priorities for the years ahead, must do so on the basis of well-known ways of living: methods that now work for at least a part of the human race. Our descendants, raised from birth with zero-gravity and adjustable seasons as commonplace elements in their lives, will be far more inventive than we are in turning those options to advantage. Those who inhabit the first communities will not have that head start; they will find enough "future shock" as it is, in making the transition from Earth to a space habitat, and for them it may be reassuring to know that they may look forward to something familiar and homelike.

In that vein, it is interesting to consider some of the possibilities for modeling directly attractive portions of the Earth. We think of a valley area two miles by twenty as rather small, but it is surprisingly large when compared to some of humankind's favorite places. Most of the island of Bermuda, including the lovely south coast areas named after the English shires, could be modeled rather well within about half the length of a space-community valley.

Only when population densities have dropped and a plentiful water source is available from the asteroids will that sort of whimsical luxury be possible. There's a small but lovely bit of the California coast, including the town of Carmel, favored by artists, writers, and many visitors. The usable area of an Island Three space community would be more than twenty-five times larger. We may expect that, in common with our ancestors who chose wistfully to call their frontier "New England," at least some of the settlers in space will model their cities and villages on the prettier areas of Old Earth.

As little as a year ago, I would have felt it necessary to write at considerable length on the structure of the habitats: the aluminum or steel cables, or the metal shells which band them and contain the forces of atmosphere and rotation; the solars which admit the sunlight while retaining the atmosphere. There is no need to do that now; a number of engineers in several government and private industrial laboratories have checked the relevant calculations. It is enough to say that the construction techniques are not basically new, being for the most part variations on the methods of Earthbound bridge building or shipbuilding. The strength assumed for aluminum corresponds to well-known alloys, with the safety factors that come out of the standard engineering handbooks. For steel cables the numbers are within normal practice for suspension bridges, higher but not double the value that was common in terrestrial bridge building as much as fifty years ago.

One problem in basic physics is worth discussing, though, because one of its possible solutions brings with it a number of diverting possibilities which the space dwellers may want to exploit. A rotating cylinder in space constitutes a gyroscope, and in the case of a space community it is an enormous gyroscope indeed. As we learned in school, a gyroscope left free will continue to point its rotation axis always in the same direction, relative to the distant stars. That is the principle on which the gyrocompass works. In the case of a space habitat, gyroscopic action could be a problem: the simple use of solar power, as well as the arrangements for natural sunlight and for the day/night cycle, all require that sunshine always arrive in a direction along the cylinder axis. One way to satisfy that condition is to orient the cylinder axis perpendicular to the community's orbit around the sun, and to provide a lightweight mirror, angled at forty-five degrees, to bring to sunshine along that axis.

Alternatively, the cylinder axis can be in the plane of the orbit. In a year, as the community moves with the Earth

M. I. T. class design for habitat
with integral shielding.

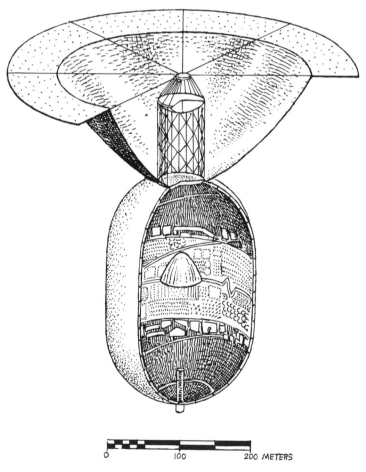

around the Sun, in that case the axis must turn through one complete rotation. In order to provide that turning motion of the cylinder axis, the motion physicists call precession, forces must be applied. The forces need not be large, because the precession rate will be very slow: only about one degree per day. Calculation shows in fact that the forces will be only about one ten-millionth of the weight that the cylinder would have at Earth's surface. Two equal and opposite forces are needed. They can be applied through hollow bearings at the cylinder ends; the bearing forces will be small by comparison with the shock loads on a locomotive wheel or on an aircraft while landing, so the bearings themselves will not be difficult to build and need not be large. At each end, the forces can be taken by tension or compression towers, as thin and spidery in appearance as terrestrial radio masts.

Where now can we obtain the lever to move this small world? One easy way is to obtain the required forces by attaching the towers to another cylinder, as nearly as possible identical in mass and size to the first. In that way each can supply the necessary forces for the other. One habitat can be located above the plane in which Earth moves around the Sun; the other, just below it. We can convince ourselves of the correctness of this solution by considering that when the two cylinders rotate in opposite directions, their net gyroscopic action will be zero. As such, there will be no resistance to turning them as a pair, and once they are established in that slow precession they will hold it to eternity without the need for more than small corrections.

For reasons of necessity, to satisfy the equations of mechanics, it seems the space dwellers may adopt a geometry that links two habitat cylinders together a form a complete community. No energy or power will be required for that arrangement, and no rocket thrusts will be necessary to obtain the necessary forces, so the solution should be inexpensive. The tension member, in fact, need be no bigger in diameter than a teacup.

If they adopt this solution, the colonists will now find that it provides them with some free benefits. The first concerns seasonal phase: the mirror schedules of the two

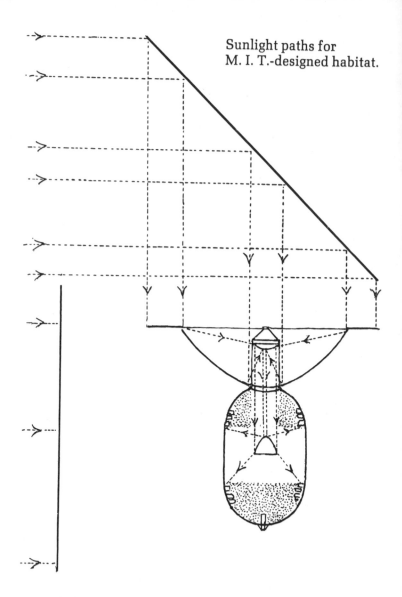

Sunlight paths for M. I. T.-designed habitat.

sister habitats will be independent, so the seasons in the two may be as different as the inhabitants wish. One may be in the midst of January weather while the other is in June. Another possibility will be to have a rather severe climate for one of the two habitats, with seasonal extremes perhaps as great as those in New England; warm summers and clear, snowy winters, for skiing and for a Dickensian "White Christmas." The other habitat, only fifty miles away, may have a climate as lush and tropical as that of Hawaii. If travel between the habitats can be arranged to be easy and inexpensive, there seem to be attractive new options for visits crossing seasons or climatic zones.

Given the two rotating cylinders in space, parallel to each other and only fifty miles apart, the space dwellers will be able to take advantage of the rotation which produces Earth-normal gravity in the habitat valleys. For Island Three, that rotation is at a rate of about four hundred miles per hour. Imagine then a simple vehicle, less complicated even than a terrestrial bus: it contains comfortably spacious seating, but requires no engine or crew. As its passengers board it, walking down stairs through their land valley as if they were entering a subway station, the vehicle will still be locked to the outer surface of the habitat. When the door is closed and sealed, a computer on the habitat will wait until the correct moment in the cylinder's rotation cycle, then will unlock the vehicle. Proceeding through space on a straight line, with the tangential velocity of the habitat, the vehicle will arrive at the other cylinder in less than eight minutes. On release it will have been given a gentle twist, so that it will perform a half-roll in the few minutes of its flight. On arrival, it will find the outer surface of the second cylinder moving at exactly its own speed, and will lock onto the outer surface in a dock similar to the one it just left. The passengers, after their few minutes of zero-gravity, will feel weight restored, and can leave their seats, take the "up" escalator, and find themselves in what is literally another world, perhaps as different from the one they left as Polynesia is from Maine in winter. Such transportation should be quite inexpensive, because it will require no energy. That sounds vaguely disquieting, rather too much like perpetual motion, but in fact the statement is true: the transfer from one cylinder to the other, in such a vehicle, will require no power. [1] Given that convenient fact, and the high utilization obtained from a vehicle that can make several trips per hour, the cost of such a journey surely will be quite low. We can imagine young people "jet-setting" over from one habitat to the other just for the afternoon, complete with their snow skis or their water skis, for no more than the cost of a bus token.

Many of Island Three's inhabitants will be commuters, going to their jobs in the cities or in the zero-gravity industries, from homes in the valley areas. With the new degrees of freedom that will exist in the space habitats, it will be possible for them to do their commuting far more comfortably and quickly than do Earth's tired millions of workday travelers.

The valleys form natural lines of communication joining the cities and their suburban areas; no village will be more than a mile from a valley center. For that short distance, bicycles or small electric-powered runabouts of bicycle speed will be quite adequate. At the valley centers, it will be natural to have rapid transit systems, but there again a new option will be possible.

Magnetic flight. Fast-moving magnet
induces image poles in conducting
surface, producing lifting force.

Within the past ten years several nations have begun investigating what is called "dynamic magnetic levitation." That is a lifting force on a vehicle which occurs when the vehicle is equipped with permanent magnets and "flies" above a conducting guideway.[2] The technology of high-field superconductors, which has been brought to commercial practicality only within the last ten years, now makes it possible for a vehicle to sustain a strong constant magnetic field without the expenditure of power. If a vehicle is standing still above a piece of aluminum, it will supply fall when released; but if it is given a forward motion, the eddy currents which its field induces in the guideway will generate counterfields, which will act always to produce lift. Dynamic magnetic levitation has several advantages as a substitute for wheels and rails in transportation systems: it is efficient, it gives a soft, gentle ride even at high speed, and above all it does not require high precision in the location and leveling of its "track." The Magnetic levitation system, sometimes called "Maglev" or the "Magneplane," is inherently capable of high speeds, from two hundred to three hundred miles per hour. On Earth there is some difficulty about attaining those speeds, because of aerodynamic drag and the high noise level produced by a train cutting through the atmosphere at sea level at so high a velocity.

In a space habitat, magnetic levitation may come into its own, because high-vacuum is an ideal dragless, noiseless medium through which a magneplane can travel at high speed. Probably the residents of space, arriving at a station within a mile of their homes, will set their electric vehicles to find their way home along the bicycle paths at a safe walking speed, following the magnetic lure of a buried wire. Entering the magneplane station, the commuters will go down through the habitat shell, and will board the vehicle when it arrives. The airliner doors will close, a diaphragm will close to seal off the entrance, and the magneplace will begin to accelerate in the high vacuum only a few feet below the valley, reaching in less than a minute a speed of three hundred miles per hour in silence. Minutes later it can decelerate for its stop at the city; or if scheduled for the zero-gravity station near the endcap's hollow bearing, it will coast on its magnetic lift up the outside of the cylinder-end hemisphere, stopping at a point where travel can continue without vehicles, in the drifting flight of zero-gravity. With unlimited low-cost electrical energy from the habitat power station, and with computer control over their movements, probably these efficient vehicles will be able to operate at intervals of only a few minutes, so that people wishing to travel to or from the cities or to the industries will not have to worry about timetables and can go whenever their own schedules make it necessary or convenient.

In an earlier chapter a method was suggested for the transport of the work force from a community to an industrial complex, which might be located at a distance of a hundred miles away; the reasons for such a location might be the need to isolate the habitat from the waste heat radiated from an intense user of energy; or the availability to an industry of workers who might choose, through personal preference, to live in habitats of different climate or architecture than those of the nearest community.

Much the same method should be usable for long-

Sunflower geometry.
Light enters at equator;
external wheel is
for agriculture.

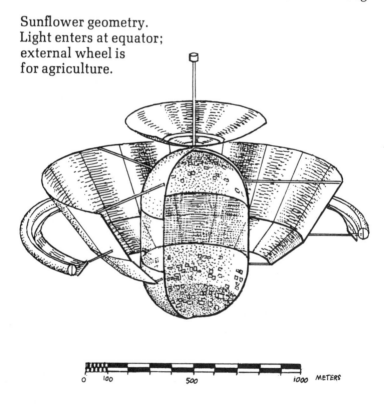

0 100   500   1000 METERS

92

distance transportation. A sphere much like the "commutersphere" I spoke of earlier, but with perhaps only half as many passengers, could provide luxurious conditions of comfort and convenience. While being accelerated the sphere could be given a rotation, leaving it with a fractional gravity to simplify such practical functions as eating or going to the toilet. I like Arthur Clarke's comment on the alternative: rapid acceleration and deceleration with zero-gravity in between: "Half the time the toilet's out of reach—the other half it's out of order."[3]

As we've learned on Earth, speeds in the jet-aircraft range are quite adequate for travel over intercontinental distances. On Earth, unfortunately, the conditions under which that travel goes on are cramped and uncomfortable: aerodynamic factors and the need for an on-board crew— to cope with weather, mechanical failures, and the tricky operation of landing—force the design of commercial aircraft too large to be intimate and too crowded to be restful.

For a flight of the same distance as New York to Los Angeles, but in an electrically-accelerated "travelsphere" plying a course between space-colonies, getting up to speed will take only a minute or so. The rest of the flight will cost nothing except that part of the vehicle's first cost and maintenance that are being amortized over that time—and, of course, the costs of food and cabin service. One very big difference between flight in the atmosphere and flight in space is that in space we don't have to worry about the speed of sound. The travelsphere can fly at higher than Concorde speeds, but with no concern about either sonic booms or pollution of the atmosphere. Making generous assumptions about the cost of the simple vehicle involved, and assuming load factors, utilization, and amortization schedules similar to those of terrestrial jetliners, it appears that flight in a travelsphere might cost about a fifth as much per passenger-mile as the ticket cost on a modern jet like the Lockhead L-1011. It's strange to think that the travelsphere would be a much simpler vehicle, but those are the facts: no engines, no complex electronics, no complicated structure to take atmospheric buffetings. And of course—no burnup of scarce petroleum resources.

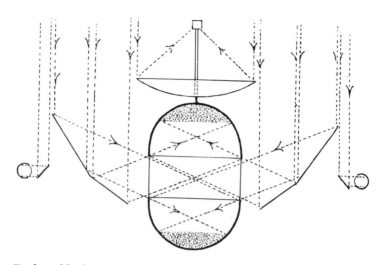

Paths of light in Sunflower geometry.

It's diverting to think what this sort of efficiency will mean for travel over shorter distances, also. In a commutersphere, for less than the round-trip cost of a short automobile trip on Earth, a man could take his wife to dinner in another community; a few minutes in the runabout, five minutes on the magneplane, a half-hour flight in a travel sphere, and the couple could be in another habitat, after choosing among dozens of them located within that travel-time of their own. That could mean selecting a concert or opera performance, or simply a dinner in a favorite restaurant, in a community which could be as different in culture and language as Rome is from Kansas City.

For good health we should spend some of our time in Earth-normal gravity; yet much of the recreation in which the residents of space indulge will surely take advantage of a new option we can never experience on Earth: to have any gravity they like, simply by riding or walking to the right distance from the cylinder axis. On the axis itself gravity will be zero, and it will increase smoothly toward Earth-normal as the valley floor is approached.

Surely new sports will be invented to make use of this

degree of freedom: three-dimensional soccer may be one example. Some old sports will also be a great deal more enjoyable in low-gravity. In a pool near the cylinder axis, a dive will be made in slow motion and the waves will break as slowly as in a dream. Those of us who enjoy scuba diving find that under Earth's oceans the need for pressure equalization reminds us, with every foot of depth change, that we are not in our natural element. A pool near the cylinder axis, or an entire sea-world, perhaps in one of the external cylinders, could have a gravity as small as a thousandth that of Earth, and could give the swimmers of the habitat the freedom to forget pressure changes and swim as naturally and freely as the fish.

It seems unlikely that any of the communities will be willing to put up with powered aircraft, because of their noise and smoke, but soaring—the use of air currents to sail in three dimensions with a glider—should be possible. As a glider pilot I find that people even on the ground seem to feel a sensation of joy and release in watching a glider fly; as Richard Bach has said, perhaps there is something of Jonathan Seagull in each of us. [4]

Human-powered aircraft concept, for use in axial region in low gravity.

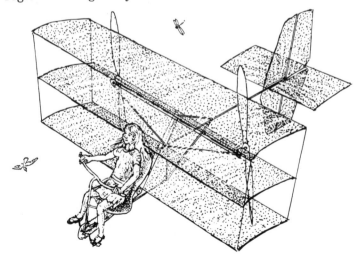

From the time of classical Greece, and perhaps even before, some men have been fascinated by the idea of flight by human power alone. Leonardo da Vinci was obsessed by it, and filled notebooks with sketches of machines which he hoped might fly. In modern times man-powered aircraft have been flown short distances, but under Earth-normal gravity human-powered flight remains an almost impossible dream. In space communities, it will become easy for everyone, not just for athletes. Close to the cylinder axes, in near-zero gravity, almost every imaginable variety of human-powered flying machine, including some of Leonardo's, will work. We can imagine elderly ladies and gentlemen taking their evening consitutionals by gently pedaling their aircraft, while viewing the world miles below them. Because they will be in a "gravity" produced by rotation, they will be able to change it at will, by flying with or against the direction the habitat is turning in. While as far from the axis as the height of a tall building, they'll be able to cancel gravity entirely by pedaling at only bicycle speed—but in the right direction.

As at swimming beaches, space dwellers may have to provide something to keep people out of danger. There are at least two possibilities: one is a near-invisible cylindrical net to prevent a tired flyer from straying too far from the cylinder axis into a high-gravity region. Another is a parachute, permanently mounted on the pedal-plane, and ready to pop open if the flyer descends too low.

Where the valleys end and the hemispherical endcap begins its upward curve toward the cylinder axis, the temptation will be great to model the mountains of Old Earth. A hike up those mountains will be a good deal easier than on Earth: as the climber makes his way to higher altitude, and starts to become tired, gravity will be lessening with every foot of height gained. By the time he's two-thirds of the way up the mountain he'll weigh only a third as much as he did on Earth or at the start of his walk, and can climb in bounding strides. At the top, two miles above the valley, he will weigh nothing at all. He will have passed the clouds at about the 3,000-foot level, so they will be far below him, but he will find that the atmosphere has lowered in density only as much as for a climb to half his elevation on the mountains of Earth.

I've devoted a good deal of this chapter to the less serious side of life in a space colony—not questions of economics and production, but of amusement and diversion. It seems appropriate to close with an account of one memorable lunchtime conversation: in the years before the topic of this book was well known, I had made a practice of challenging skeptics to name their favorite sports, and then always pointing out that the sport could be done better in space than on Earth. Finally someone named a delightful sport that, even in these uninhibited days, is carried on only in private. The skeptic instantly became a believer: can one imagine a better location for a honeymoon hotel than the zero-gravity region of a space community?

7
RISKS AND DANGERS

Almost every human activity carries with it some element of risk. Occasionally, in a rare macabre frame of mind, I have reflected on the fact that at any time almost every human being, however healthy, is within one or two minutes of death if the wrong combination of circumstances were to come to pass. When I lecture on the topic of habitats in space, it is natural that some of the questions that follow relate to the possibility of violent catastrophe in a space community. Given the fragility of life, that possibility will always be there, so we must be quantitative and estimate the risks that will attend the human settlement of space. It is reassuring to find that in fact they are rather less than those to which we are exposed every day here on Earth.

Almost invariably the first question that is asked about space habitats concerns meteoroids. These are, for the most part, grains of dust which have been in the solar system since its formation several billion years ago. As our Earth revolves around the Sun each year we travel at a near-constant speed of about thirty kilometers per second: higher than any of the relative speeds needed for launching a satellite or traveling to L5, or even for voyaging to an asteroid. Most of the grains of dust which we encounter in our annual passage around the Sun are moving relatively slowly, so typical relative speeds with which we meet them are just our own. Almost the highest-speed meteoroid which has ever been measured corresponds to a dust grain moving in a circular orbit around the Sun, but in a direction contrary to our own; combined with our own velocity that gives an encounter at doubled speed.

Most of these meteoroids are of cometary rather than asteroidal origin, and can be thought of as dust conglomerates, possibly bound by frozen gases.[1] If present scientific ideas are correct, therefore, a typical meteoroid is more like a minisnowball than like a rock. Even a very small meteoroid carries, because of its velocity, a great deal of energy, but fortunately almost all meteoroids are of microscopic size: in the frequency curve of their occurrence, as the size increases the number goes down rapidly. Spacecraft sensors have collected abundant and consistent data on meteoroids in the range from one gram (that is about one thirtieth of an ounce) down to a millionth of a gram.[2] Above that size, there is so small a chance of finding a meteoroid that even in a voyage of years a spacecraft records almost no data.

For relatively large meteoroids, the series of Apollo flights has left us with a scientific legacy especially important for just this question: the Apollo seismic network, a series of very delicate seismometers left on the Moon. These instruments continued to record for many months after the flights which installed them, and they have recorded not only Moonquakes but the collisions of meteoroids with the lunar surface. So sensitive are these machines that their builders claim to be able to detect every strike occurring anywhere on the Moon by a meteoroid of soccer-ball size or larger. Fortunately these two independent means for measurement of the meteoroid-size distribution agree quite well, and allow us to estimate with some accuracy the chance of a strike on a space habitat, for a meteoroid of any given size.

There is a third method for the measurement of meteoroid-size distribution. It is ingenious and relatively inexpensive: an array of wide-angle cameras, forming a pattern which is called the "Prairie Network" is distributed over about one million square miles of lightly populated farming states in the central part of the United States. When a meteoroid enters our atmosphere, leaving the luminous trail which we call a meteor, the Prairie Network sky-cameras photograph the trail with such accuracy in space and time that the position, altitude, and velocity of the meteor can then be calculated. Some of the best measurements of speed distributions come from data of this kind.[3] Unfortunately, it is much harder to obtain from that source accurate figures on size distributions. Those have to be based on the brightness of the trails observed, and then on a crucial assumption: how much of the energy of the incoming meteoroid is converted to heat and light.

The Prairie Network data agree with those of the other two methods quite well for meteoroids the size of a marble. They aren't in such good agreement for the larger or smaller ones, probably because of the assumptions made about luminous efficiency. If one assumes, as is consistent with the most common modern view, that the typical meteoroid is a dust-conglomerate, then the efficiency of conversion of the incoming energy to heat and light should be rather high. With that assumption the camera data agree better with those of the other two methods than they do if a low efficiency is assumed.

Averaging the data from what seem to be the most reliable sources, one finds that in order to be struck by a meteoroid of really large size, one ton, a large "Island Three" community would have to wait about a million years. Such a strike should by no means destroy a well-designed habitat, but it would certainly produce a hole and cause local damage.

In order to find meteoroids that would strike at a frequency high enough to worry about, we have to consider much smaller sizes, of about the weight of a tennis ball. On one of the big communities, there'd be a strike by one of those about every three years. Curiously, there is a reason why a habitat of given size would be struck less often than an equal area at the top of Earth's atmosphere: the gravitation of Earth is so strong that it "sweeps out" meteoroids, sucking them in from a region of space much larger than its own area. The space habitats, far enough away from Earth not to be in the affected region, and having almost no gravity of their own, would be stuck relatively less often.

The most vulnerable parts of a habitat will be its windows; they will occupy a large area and, being made of glass, will be relatively fragile. They will naturally be subdivided into small panels, for two reasons: to guard against the possibility of catastrophic damage, and to allow the aluminum, steel, or titanium supporting structure to carry all the structural strength in the window regions. A window panel may have an area two or three times that of a window on a jet aircraft. With such a size, the metal frames that carry all the structural loads can be so thin that they will be invisible from a valley floor, and the windows will appear continuous when viewed from such a distance.

For panels of that size, the loss of one will certainly not be catastrophic for the community. For what we have called "Island Three," if one panel were blown out entirely it would be several years before the atmosphere would leak out. Detection of a blowout should be almost instantaneous: it would result in a plume of white water vapor, condensing to ice crystals in vacuum, visible from the sister habitat. If a patch were put on the blown out panel within an hour, the loss of water vapor would be economically tolerable (the oxygen would cost far less to replace) and probably no one but the repair crew would even know of the event.

Even for the smallest community, Island One, the corresponding numbers would be quite tolerable. There it would be several thousand years between strikes by a meteoroid big enough to break a window panel. When a panel blew, if it were patched within an hour the loss of atmosphere would reduce the pressure by only about as much as we would find on Earth in climbing a hill two hundred feet high: not even enough for us to detect a pressure change on our eardrums. For the most recent design of Island One, these risks would be further reduced by a large factor. We now assume a design in which heavy shielding, provided for cosmic-ray protection, would protect the window areas from any direct "view" of space.

At the surface of the earth we are exposed to radiation from three different sources: emanations from the soil, rocks, bricks, and other structures which make up our environment, radiation from small quantities of radioactive substances within our own bodies, and cosmic rays which penetrate our atmosphere. Radiation is measured in units of Roentgens, and for biological damage the unit rem (roentgen equivalent man) takes account of the differing amounts of damage done by radiations of various kinds. For total dosage over a period of time, the unit is the rad (radiation dose). On Earth's surface the amount of radiation to which people are exposed varies over an enormous range, depending on where they live.

Oddly enough, most of the radiation the average person gets comes from inside: trace amounts of radioactive elements in the body. The radiation from outside depends on such details as whether one lives in a brick house (bad) or a wooden house (good). Most of all, though, it depends on geographical area; in the monazite-sands region of India the residents get a natural dose of almost one rad per year.[4]

103

By comparison, our normal dose from cosmic rays is relatively small: least of all at sea-level near the equator, but still only a small fraction of a rad per year for a mountain elevation in a temperate latitude. At the poles it is much higher; the latitude differences arise from the fact that Earth possesses a magnetic field which provides it with a substantial amount of protection against the lower-energy cosmic rays.

When all the sources of natural radiation, internal, external, and cosmic, are added, they amount to an average dose of about a third of a rad per year for a typical Earth-dweller. After a great deal of testing and years of discussion, to which many physicists and biologists contributed, the Atomic Energy Commission (in the days long before it was called ERDA) settled on an allowable annual dose for its workers of five rad per year, and of a tenth that for the total U.S. population.

Clinically, only the most sensitive and delicate laboratory tests can detect effects in humans from average radiation of less than about twenty rad per year, and far larger average exposures are required before a human individual is aware of any consequent illness or discomfort.

In space, far from the protective shield of Earth's magnetic field, the level of steady, highly penetrating cosmic rays (the so-called primary galactic radiation) is about ten rad per year. If there were no other radiation to consider, it would be reasonable to consider building the first space habitats with no shielding at all.

If a large fraction of the world population were to live in those conditions for many centuries, we should be concerned about the resulting increase not only in cancer but in the rate of mutations. That would not occur, though: the buildup in the size of habitats to the point of thorough shielding would take place over at most a few decades of time, and during that brief time only a small segment of the human population would be exposed to enhanced radiation levels.

There is however a more serious cosmic-ray problem, arising from a type of radiation to which we are never exposed on Earth: these rays are the "heavy primaries": nuclei of helium, carbon, iron, and the whole range of elements found on Earth. They form only a tiny fraction of the total cosmic radiation, but they are far more damaging than the rest.

When heavy primary cosmic rays pass through material, they leave a dense trail of ionized atoms. These atoms are highly active chemically, and are so numerous that in living cells they cause cell death. The same property of intense ionizing power, which is responsible for the biological damage done by heavy primaries, is also a protection against them: in our atmosphere they lose energy so quickly by ionization that they are absorbed at high altitudes, never penetrating to sea level.

The only direct human experience with heavy primaries has been that of the Apollo astronauts, who ventured outside not only the atmosphere but also the protective magnetic shield of Earth. In that open region they observed flashes of light, visible especially when they adapted their eyes to total darkness. Most scientists who have studied the subject agree that these light-flashes were almost certainly caused by heavy primaries. On *Apollo 17* a systematic study was made of this effect. When I asked Dr. Harrison (Jack) Schmitt, who went to the Moon as an Apollo 17 scientist-astronaut (and later was elected U. S. Senator from New Mexico) about his observations, he reported an odd fact: although the light-flashes were visible at a rate of one every few minutes throughout most of the voyage, during the period of one deliberate experiment none were seen for an interval of an hour or so; at present no one has come up with a good explanation of how they could have vanished, even temporarily.

On *Apollo 12* the astronauts were exposed to the heavy primaries for about two weeks. Estimates based on direct radiation measurements and the known sizes of body cells suggest that during that period their loss of brain cells was a few in a million; a similar figure holds for retinal cells, and for the very largest body cells (neurons) the fraction is perhaps as much as one in ten thousand.[5] These are small numbers, but there is still reason for concern about them: the cells involved are nerve cells, and as such are not replaced by the normal body repair mechanisms. We have then one "data point" which we could take as conservative for our further calculations: the *Apollo 12* crew was exposed to a certain known dose of the heavy primaries, and suffered no apparent ill effects from them. To be on the safe side, therefore, our design of even the first space habitat should be based on the requirement that in a working career of several decades a human being would be exposed there to a total dose no greater than that which was received in only two weeks by the *Apollo 12* astronauts.

Occasionally, for reasons we are only slowly coming to understand, the Sun emits sudden bursts of radiation called flares. These rays travel almost as fast as light, and reach Earth within minutes. When they do, they cause brilliant auroral displays in the upper reaches of our atmosphere. Very rarely, every few decades, particularly intense flares occur, which saturate Earth with radiation, temporarily blank out much of our long-distance radio communications, and even affect Earth's magnetic field. Such an event last occurred in the 1950s. If there had been astronauts on their way to the moon at that time, they would almost surely have been killed by that flare. Therefore, even the first space community must be protected against solar flares and heavy primaries. This could be done by passive shielding, using lunar surface material or the slag from the industries of the early colonies. The thickness required would be some fifty centimeters (twenty inches) of sand or its equivalent. That would be enough to increase noticeably the required mass of Island One.

The effect of that shield thickness, oddly enough, would be to enhance to an unacceptable level the radiation from the galactic primary rays. The reason is that on encountering dense matter those particles would break up into many more, of lower average energy but much greater total numbers.

In the end, then, one must do the entire job and get rid of all three components of radiation. When the numbers are

Shielded "Model One," an early habitat design with farms in cylinder valleys, living areas in shielded end-caps.

106

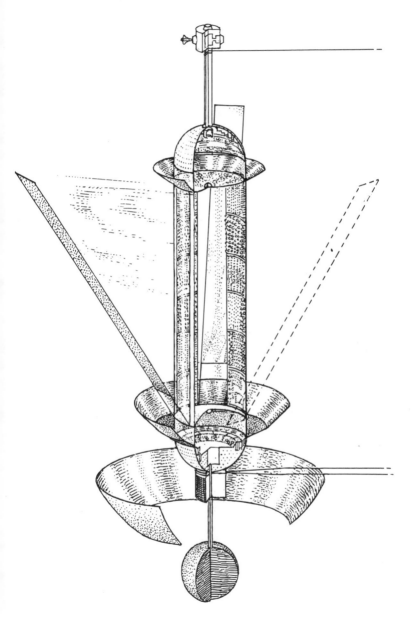

worked out, one finds that the shielding needed is substantial: equivalent to about two meters (over six feet) of soil. Once that problem is thoroughly understood, it constitutes a serious restriction on the design of the first habitats. Fortunately, a geometry has been found that fully satisfies even the most severe shielding requirement, without sacrifice of desirable design features.

The later space communities, of the size of Island Three or larger, will have atmospheric depths, and thicknesses of structure below the ground, so great that they too will afford to their inhabitants protection from cosmic rays comparable to that of Earth. Their building materials, the lunar soils, are already known to be fairly similar to those of Earth in natural radioactivity.[6]

To summarize, with proper design both the early and the later space communities can be shielded against all types of radiation, to levels comparable with what is found here at the surface of the earth.

In order to minimize costs, probably the early habitats will have atmospheres composed mainly of the material most plentiful on the Moon: oxygen. The National Aeronautics and Space Administration has reason, though, to be apprehensive about pure oxygen atmospheres. In 1967 three prospective Apollo astronauts died in a flash fire in an Apollo module at Cape Kennedy, during a test conducted in pure oxygen.

The conditions of a space community will be different in several ways. First, the oxygen pressure will be only one-fifth as high. At the Cape in 1967 the disastrous test was conducted with oxygen at the full sea-level pressure that's normally made up mostly of inert nitrogen. Second, the volume of a habitat will be millions of times larger than that of an Apollo module, so that any small fire which starts within it cannot build up the gas pressures which were destructive in the Apollo test.

Possibly, though, these two differences will not be enough. To be on the safe side we want an additional security factor. One approach is to add a special component to the atmosphere, something that is harmless to humans but that either would not support combustion, or

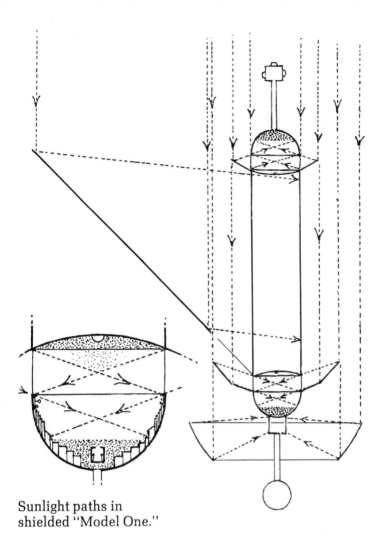

Sunlight paths in
shielded "Model One."

would actively damp it. We should first consider obtaining
a damping gas from lunar materials. On Earth, fires are
partially damped by the presence in our atmosphere of
nitrogen. The lunar surface materials are known to contain
small amounts of volatile gases, so that in processing a
million tons of lunar materials a few thousand tons of
gases will be evolved. Their composition is not as accurately known as we would like, but it is thought to be

mainly carbon dioxide, nitrous oxide and a small percentage of water. We might be able to get a useful amount of nitrogen from that source. It does not seem likely, however, that nitrogen will be a very effective fire retardant. Even if we find a cheap source for adequate quantities of it (which seems unlikely), we cannot put much nitrogen into the space-community atmosphere without raising the pressure enough to increase the habitat structural requirements.

There are gases which are harmless to humans at least for short times, but which actively retard fires; some of the freons have this property. But these are chemicals made of elements not all of which are found on the moon, and we lack adequate data on their long-term physiological effects.

It appears now that the simplest solution would also be the best. To maximize the day-to-day pleasures of life in the space colonies, as well as their safety, it seems wisest to bring along from Earth enough hydrogen so that the atmosphere will have a comfortable relative humidity, and so that there will be plenty of lush green vegetation. Structures there will be made of nonburning materials, similar to brick or cinder block on Earth, so with a combination of reduced atmospheric pressure, large total volume, and plenty of water the fire danger appears reducible to an acceptable level. This is an area in which actual laboratory research here on Earth will be required before the answers are certain.

With regard to war we must be speculative. I hesitate to claim for the humanization of space the ability to solve one of mankind's oldest and most agonizing problems: the pain and destruction caused by territorial wars. Cynics are sure that mankind will always choose savagery even when territorial pressures are much reduced. Certainly the maniacal wars of conquest have not been basically territorial. When Genghis Khan conquered most of Europe and Asia he had no plan in mind for the conquered lands, and therefore simply destroyed their cities and murdered their people. Yet the history of the years since the second world war suggests some changes relative to the past; if anything, that warfare in the nuclear age is strongly, although not wholly, motivated by territorial conflicts: battles over limited, nonextendable pieces of land. It appears that the territorial drive to conquer someone else's land should be muted under the conditions of the space communities: they will be free of the age-old associations which fuel territorial wars on Earth, they will be replicable so that no one need feel constrained by a fixed boundary, they will be independent of each other for their essential needs, and they will be movable. In the long run, when new habitats may be built most economically at the asteroids themselves, upon completion their residents will have a choice: to move, by low-thrust engines over a period of decades, to an area in which other, culturally congenial communities are already located, or—to go the other way.

From the viewpoint of international arms control, two reasons for hope come to mind. We already have an international treaty banning nuclear weapons from space, and the space communities can obtain all the energy they could ever need from clean solar power. The temptations presented by nuclear-reactor by-products need never exist in space.

From the viewpoint of a military man, the space habitats will seem rather unpromising as sites for weapons or military bases. First of all they will be quite vulnerable militarily, so that no one in such a habitat can be tempted into believing that he can attack someone else without risk to himself. Second, their distance from Earth, and their consequent separation from it by at least one or two days of travel time, will mean that they can never be used as effective sites for an attack on the home planet. In summary, the probability of wars between the habitats seems, to me at least, considerably smaller than that of wars between nations on Earth.

At lectures on space communities, an occasional question concerns the possibility of attack on a colony from within, by some insane person or extremist group bent on mutual annihilation. The possibility is there, at some level, but probably it will carry with it some safeguards of its own. I suspect that many habitats may choose to have some sort of "customs inspection" which would eliminate or greatly reduce the likelihood that explosives or weapons could be introduced into them. In the past years on Earth we have come to take inspections of this kind as a matter of course at all airports. If, in spite of such precautions, a terrorist were somehow to import or manufacture explosives, he would have to do so on a fairly large scale to produce a major catastrophe. Like airplanes, bridges, and ships, the habitats will be designed so that loss of a single supporting band, or of a single longitudinal cable, will not result in a major rupture but only in the redistribution of loads to the supporting members nearby. As discussed earlier, the destruction of one or even several window panels would result only in a loss of atmosphere slow enough that there would be plenty of time for evacuation to communities nearby.

The external tension and compression towers, which may provide for each cylinder the forces necessary for its precession about the Sun, would not be very vulnerable to terrorists, located as they would be in space where no one could move without a space suit. If, though, one of them were to be destroyed, either by accident or by intention, it wouldn't result in catastrophe to the habitat. The precession would be arrested, so if repairs took as much as a day the residents would see the image of the Sun's disc wobbling by about two solar diameters, though the intensity of sunlight would be undiminished. On completion of repairs the precession rate could be speeded up to a rate greater than normal, until the community "caught up" to the correct orientation. Such an event would be seriously damaging only if repairs took more than one or two weeks, so that Sun angles were changed by many degrees and crop growth were correspondingly affected.

Certain dangers exist on Earth but would not in a space habitat; earthquakes and volcanoes are among these. Often they wipe out thousands of people at a time, particularly in seacoast areas. Tornadoes, hurricanes, and typhoons also kill, and numbers of people are killed every year in small-boat accidents through weather or violent waves. Among the risks which our technical society has added are those of automobile accidents. Because of good roads, safe automobiles, and relatively strict traffic laws, in the United

112

States we have about the lowest accident rate per passenger-mile that is found anywhere in the world; yet even our rate results in the death of 50,000 people per year, out of a population of two hundred million. One comparison between the risks on Earth and those in a space habitat is instructive: even in the extreme case in which it is assumed that a meteoroid strike of one-ton size on a space habitat would result in total destruction and the loss of all the inhabitants, the risk of death from that cause would be only one sixtieth of that which we run in the United States by the existence of our automobiles.

If the space-habitat option is followed on the earliest possible time scale, the result could be that within a few decades the nations of the world would all be dependent on solar energy from satellite solar power stations built at space communities. Nuclear energy, under those conditions, would be confined mainly to the laboratory. Dependence on a relatively vulnerable but inexhaustible power source would remove one of our present causes of international tension and the threat of war, and at the same time would deter any would-be adventurer-nation from carrying out an attack on a neighbor.

In contrast, if for our energy we are forced to rely on a rapid, large-scale development of liquid-metal fast-breeder reactors, within a few decades every industrial nation and every developing nation will have such devices. Plutonium will be in production in large quantities in every such nation, and the temptation to divert it to weapons production will be very strong for at least some political leaders. With so much fissionable material being produced and shipped, it seems likely indeed that some of it will be diverted by terrorist groups, and consequently Earth may become a much more dangerous place than it is now. [7]

In terms of risk, therefore, the alternative appears to lie between a development of space communities, relatively safe from catastrophe, in which an increasing fraction of the human race would be widely dispersed and consequently safe from simultaneous destruction, and an Earth ever more crowded with population, on a strictly limited land area, under conditions in which the probabilities both of war and of terrorist acts would be enhanced.

8
THE FIRST NEW WORLD

The first space community large enough to form a powerful industrial base, able to manufacture products of value
in quantities great enough to provide important economic
benefits to Earth, will require a population of at least
several thousand people. A space-station supporting only
a few astronauts would be far too small to "seed" the

manufacturing program. To build a sizable habitat will require making full use of the advantages of scale. The experience of space exploration so far is that the development costs of new vehicle systems tend to be under-estimated, while the economies of scale, of quantity, and of size tend to be insufficiently used. Beyond a certain lower limit, the cost of greater tonnage transported to orbit is only that of additional launch operations, which become less expensive as once-developed systems are replicated and progress is made along the learning curve for their construction. For that reason it wouldn't cost ten times as much to establish in space a work force ten times as large. We can't say for certain what will be the minimum number of people needed in space in order to reach the "ignition point"—the level where they will be generating new wealth fast enough so that further growth won't require subsidy from the Earth—but all the studies made so far agree that ignition will be reached by the time the popula-tion in space reaches 10,000. If those people are only as productive as an equal number engaged in heavy industry on Earth, their output every year of finished products will be more than the mass of several ocean liners.

Concentrating on the "nuts and bolts" details of the construction of Island One, we must keep clearly in mind the difference between science fiction and reality: the difference is the contrast between practical technology and unchecked imagination. We must depend only on present-day technology, on machines which we are sure we can build within the limits of our present knowledge, and on costs calculated with as much realism as we can attain. Time scale is of the greatest importance. Unless Island One can be built rather quickly, its productivity will be of no use to us in the time scale for which we may be willing to commit investment. That constrains us to what the professionals call "near-term" launch vehicle systems. In our design work we must restrict ourselves correspond-ingly to practical engineering and sensible economics.

In the early days of every remote construction project, accommodations are modest and living conditions are rather simple; the amenities come later. We've seen that

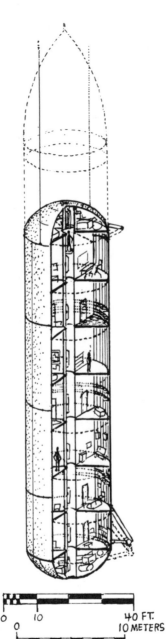

A3

Modular habitat
using shuttle
external tank,
for early rapid
buildup of space
manufacturing.

0   10   40 FT.
0   10 METERS

history repeated with the construction of the transcontinental railway in the last century, and with the opening of the Arabian oil fields during the past few decades. By the time the population in space passes the ignition point, though, we can expect that pressures will be strong to transfer from modular apartment-like habitats to something large and Earthlike. I'll give now an "existence proof," a demonstration that one possible, workable design exists to house and support 10,000 people in comfort and safety. No one will be more surprised than I if, when Island One is completed, it looks very much like the sketches we now make of it. Even its size and its population may be quite different. If we go by the almost universal human experience of large-scale construction projects, it will probably end up to be smaller, and will cost more, than our first estimates indicate. Knowing that from the start, we should take care to develop the design of Island One in such a way that it can be reduced in dimensions, or as engineers say, "de-scoped" as the design progresses.

For conservatively chosen figures on agricultural productivity, we'll need a growing area about equal to a square 0.8 kilometers on a side. There will be no need for that growing area to be spacious or beautiful; an agricultural plant does not care whether it has an open sky above, or only a ceiling. Sunshine in great quantity will be required, as will water, soil, and nitrates.

Plants are relatively insensitive to radiation, so there appears to be no need to provide the agricultural areas with radiation protection. In the early days, though, before we have sufficient experience, it may be wise to grow our seed crops within the living-habitat where full protection from cosmic rays and solar flares will be provided.

One quite efficient design for agricultural areas consists of a series of partial wheels (tori) connected to form large fields all at the same level. Planting and harvesting machines as large as the biggest combines ever seen in the wheat-fields of the western plains can move freely over those fields without obstruction. Sunshine will enter through glass windows, and the appearance will be not unlike that of a large greenhouse. In comparison with the alternatives, this design will use so little structural mass that the efficiency of agricultural productivity will become unimportant: if after additional research it is found necessary to double the area allocated to agriculture, that change will add very little to the total structural mass of Island One.

More than a century ago Queen Victoria's consort, Prince Albert, led a distinguished group of British industrialists in the design and realization of the International Exhibition of 1851. The central feature of the exhibition was the Crystal Palace, a light and airy structure made of glass windows set in modular ironwork. So light and so well-designed was the Crystal Palace that it was assembled within a few months, by a construction crew of quite moderate size. A whole avenue of trees and acres of exhibition space were enclosed. Our multiple-torus geometry for agricultural areas strikingly resembles the Crystal Palace, with its arching vaults of glass.

As in the case of high-yield agriculture on Earth, most

Multiple-wheel "Crystal Palace" with cables to rotation axis gives maximum sunlit normal-gravity area for lowest mass.

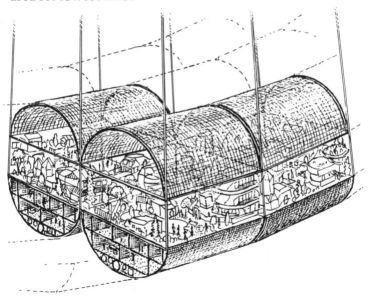

farming activities will be mechanized, so radiation shielding for the operators of tractors and combines can be incorporated into the machines themselves.

Light industry, of bench-top scale, may be carried on within the living-habitat, but heavy industry can make use of the zero-gravity of free space.

The design requirements for the living habitat in Island One are severe. The habitat must admit sunshine easily, yet be fully shielded from cosmic radiation. It must provide a spacious, comfortable environment, with long sightlines to prevent the inhabitants' suffering from claustrophobia. Ideally, it should provide easy access to a region, fully shielded, where zero-gravity sports can be enjoyed. For safety, mechanized transport should not be relied on: in the event of a sudden major emergency, it should be possible for the entire population to move rapidly, without mechanical assistance, to docking ports for evacuation. Finally, the habitat must be economical of mass, both in structure and shielding.

The land area of a habitat for 10,000 people can be estimated from considerations of "personal space" and the experience of small towns on Earth. A typical garden-apartment community in an affluent section of the United States provides, with its swimming pools, tennis courts, and landscaping, about 45m² of total land area per person. For comparison, the city of San Francisco, averaging over both residential and park areas, provides about twice that area for its population. Some of the attractive hill towns in southern France and Italy have only about one-fifth as much.

One possible geometry that satisfies all these requirements is simple and structurally strong: a sphere one mile in circumference, with sunshine brought inside through windows. If the sphere rotates twice per minute, it will provide Earth-normal gravity at its equator, near which most apartment areas can be located. At the forty-five-degree "lines of latitude" halfway up the inner surface of the sphere from the equator, gravity will be a third less than Earth-normal. That variation from Earth conditions may be our self-imposed "design limit" until we gain experience on physiological tolerances.

In such an environment each family of five people can enjoy a private apartment as large as a spacious house (230 square meters of floor area) with a private, sunlit garden of a quarter that area. By arranging the apartments in terrace fashion, only a small fraction of the total spherical surface area below forty-five-degree latitude need be devoted to apartment gardens, most of the remainder being available for parks, shops, small groves of trees, streams, and other areas available to all inhabitants.

The sunshine will enter, during a day-length set by the settlers' choice, always at a fixed angle. That will permit providing every room of every apartment with natural sunshine throughout the day. On Earth, a narrow aperture between buildings can receive sunshine only for a few minutes each day, but not so in space, where each window may look out onto a sunlit, private mini-garden.

The equator seems an ideal location for a wandering, shallow river, opening into occasional deep pools for swimming. The shoreline beaches can be of lunar sand,

Multiple wheel geometry. Used in
Crystal Palace for living and farming
areas, and in Island One farms.

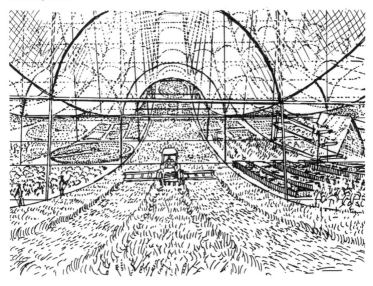

121

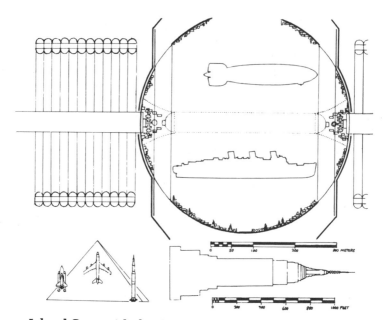

Island One, with the *Queen Mary,*
Hindenburg, Empire State Building,
Saturn 5, and Great Pyramid for scale.

and perhaps at a little distance, surrounded by greenery, there can be paths for bicycling, walking, and running.

When structural details are examined, it develops that the optimum location for windows will be near the rotation axis. There, only the pressure load will be important, and gravity will add little to the structural demands. The sphere will be no fragile eggshell, though. Its aluminum wall will equal the thickness of battleship armor, up to seven inches at the equator.

Low-gravity swimming-pools and "hangars" for human-powered aircraft can be located near the rotation axis. Walking to them from the equator will be equivalent to climbing a gentle hill, and should take only about twenty minutes.

For a given volume enclosed, a sphere is the shape that requires the least surface area. That is important for minimizing the required mass of cosmic-ray shielding. For

122

economy, the shield can be made of unworked lunar soil or industrial slag packed between thin spherical shells spaced a few meters away from the rotating habitat. It is possible in such a geometry to bring natural sunshine into the habitat through mirrors all of which are stationary in space. Only much later in the history of space communities need the designers concern themselves with such complications as rotating mirrors.

With complete shielding, provision must be made to remove from the living habitat the heat brought inside by sunlight. One easy way appears to be through large axial passageways divided by a cylindrical shell. Air circulation through these passageways will remove the heat to external radiators, and the same corridors will serve for the zero-gravity movement of people and freight to and from the industries and docks outside.

If desirable, it will be rather easy to separate the sphere visually into three "villages." That arrangement will permit making the day-length and time of day of each village independent of all the others. That in turn will allow a convenience and source of efficiency forever denied us on Earth: In order to get the most out of machines, chemical processing plants, and other industrial facilities, they should be run full time. On Earth, in order to do that we must subject people to working night shifts, which almost no one likes. In Island One, though, three villages can run at time zones separated by intervals of eight hours, so that industries can run full time while everyone remains on his own "day shift."

For structural simplicity, we want to avoid in our design any rotating pressure-seals; the habitat should rotate as a unit, airflow being contained within a single pressure vessel. Combining the Crystal Palace geometry for the agricultural areas with a central sphere for people, we arrive at the design concept called Island One.

The structural mass of Island One has been checked by calculations in several studies, and is about equal to that of a large ocean liner like the *Queen Elizabeth II*, 100,000 tons. Buildings, soil, and atmosphere will be several times as much, and even in this most efficient design shielding will add another three million tons.

To summarize, Island One will be small, though far less crowded than many Earth cities, and it can be attractive to live in. The inhabitants can have apartments which will be palatial by the standards of most of the world. Each apartment will have a private garden, bathed every day in sunshine at an angle which will correspond to late morning. Even within the limits of Island One and its water supply the colonists can have beaches and a river, quite large enough for swimming and canoeing. The river will offer a possibility that some people will be sure to exploit: a float trip, past the dam, filters, pump area, and spillway which interrupts the circular river at one point, all the way around the cylinder circumference to the starting point.

Even within Island One the new options of human-powered flight and of low-gravity swimming and diving will be possible, and the general impression one will receive from a village will be of greenery, trees, and luxuriant flowers, enhanced if the village chooses to run with the climate and plantlife of Hawaii. Heavy industry can be located outside but nearby, so that no vehicle faster than a bicycle will be needed throughout the community. Island One will rotate about once every thirty-one seconds, to provide Earth-normal-gravity for its inhabitants whenever they are at home. Only when at work outside the habitat will the residents be subjected to zero-gravity; in a daily routine of that kind their bodies will retain normal muscle tone and strength without special exercise.

The site of Island One should be far enough from Earth and Moon to avoid frequent eclipses, so the community can use free solar power continuously. We don't want it so far from the Earth that transport will be difficult, nor as close as the Van Allen radiation belts that surround the Earth. When all the logistics are considered, the best location may be simply a high circular orbit, with a period of a few days, partway out toward the Moon. There is another choice, attractive mathematically, which was studied intensively for a time: it is an eccentric orbit with a period of two weeks, just half that of the Moon. Still earlier, those of us interested in answering the question "Where will the colony be?" had considered a point on the

124

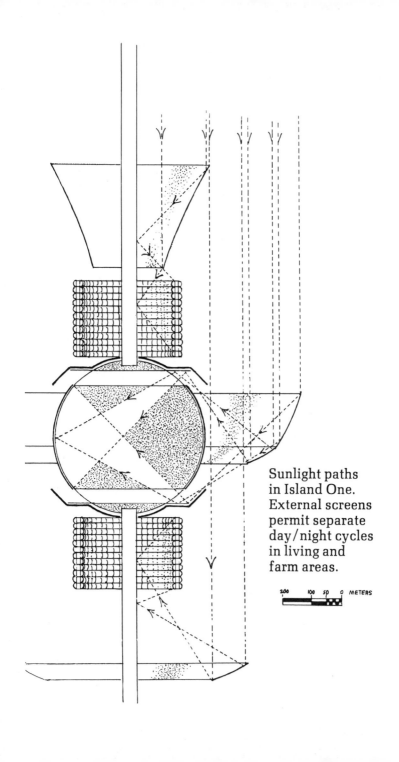

Sunlight paths
in Island One.
External screens
permit separate
day/night cycles
in living and
farm areas.

200   100   50   0   METERS

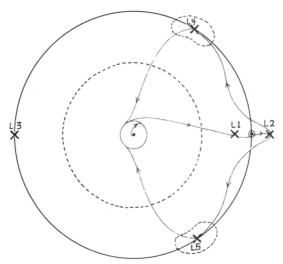

The five Lagrange points of the Earth-Moon
system. Inner circle is geosynchronous
orbit; dashed circle is possible colony orbit.

Moon's orbit called L5, the fifth of several locations in
space whose properties were first described by the
French-Italian mathematician and physicist Joseph Louis
Lagrange (1736-1813). In the language of the 1911 *Ency-
clopaedia Britannica*, "He gave proof of the undiminished
vigor of his powers by carrying off, in 1764, the prize
offered by the Paris Academy of Sciences for the best essay
on the libration of the Moon.

"His success encouraged the Academy to propose, in
1766, as a theme for competition, the hitherto unattempted
theory of the Jovian system. The prize was again awarded
to Lagrange, and he earned the same distinction with
essays on the problem of three bodies in 1772, on the
secular equation of the Moon in 1774, and in 1778 on the
theory of cometary perturbations."

Lagrange used the gravitational theory developed by
Newton to explore the special properties of two unique
points in the orbit of Jupiter. One of these points preceded
the planet in its orbit around the Sun by sixty degrees,
while the other followed by the same amount. Lagrange
concluded that these were in fact stable points, near which

126

any objects with the correct initial location and velocity would stay forever. From that time on these were known as the fourth and fifth Lagrange-points, described by solutions to what physicists call the restricted three-body problem. Years later, observations through primitive telescopes showed that several asteroids or minor planets were trapped near the Lagrange points. These became known as the "Trojan" asteroids.

If we want to use the corresponding Lagrange-points in the Earth-Moon system either as sites for colonies or as possible locations for useful trapped materials, we are up against a much tougher kind of mathematics. We must solve not just a three- but a four-body problem, because the Sun, distant as it is, powerfully affects orbits in the vicinity of Earth, in consequence of its enormous mass.

Fortunately, the problem has been done for us, but only just in time. In 1970, A. A. Kamel, a student of Professor John Breakwell at Stanford, obtained his doctor's degree in engineering by publishing a thesis with the forbidding title "Perturbation Theory Based on Lie Transforms and Its Application to the Stability of Motion Near Sun-Perturbed Earth-Moon Triangular Libration Points." Dr. Kamel's work, which gives us in an elegant mathematical form a solution that had already been obtained by the brute-force methods of computer calculation, tells us that in the Earth-Moon system L4 and L5 are no longer stable points, but that they are replaced by something at least as good: stable regions which move in orbits of very large dimensions about L4 and L5, on a slow, eighty-nine-day cycle. The properties of L4 and L5 are so unique that a society has been named after L5, and for convenience we often speak of "L5" as a nickname for "any orbit above the Earth's radiation belts, and no farther away than the Moon." It's characteristic of the orbit-mechanics problem that the experts in that field often rush in, waving great stacks of computer output, and lecture the rest of us on a newly discovered orbit that's better than any found before. By now this has happened enough times that a wise man wouldn't place bets on exactly where Island One will go. The one clear message we can be sure of is that there's room in high orbit for a total population many times that of the Earth. We need not fear, by the way, that eventually our local neighborhood, the Earth-Moon system, is going to become overcrowded. Space communities could be located on orbits almost anywhere in our solar system, and with proper mirror design could still enjoy the same solar intensity which we have (on good days) here on Earth.

We can set a scale for the investment required in Island One by considering the largest space project we have so far carried out: Apollo. That venture, which will surely be remembered long after the misery and the horrors of our century have been decently laid to rest in the history books, cost about $50 billion in money of 1978 vintage, halfway between the "Apollo years" of the 1960s, and the late 1980s, which conceivably could be the years of Island One. Apollo was begun at a time when the mood of the nation was vastly different from what it is now: then we had confidence in our abilities, we saw our living standards increasing rapidly, our money was sound and we did not yet see limits to our continued growth. Though the environment was deteriorating as a result of our industries and our transport systems, most of us were unaware of the fact.

Now these positive factors are reversed. The late 1960s and 1970s have become a time of disillusion, of slow economic growth coupled with inflation, and of living standards improving only slowly. Soon after the first Apollo landing, in 1969, we passed through a period of profound distrust of anything technological, and we will probably never again welcome new technical options in the same unthinking manner that we did in the 1950s. This is probably all to the good. Our physical power has grown so much that we should now examine with the greatest care and considerable cynicism any new technical proposal, lest it carry with it unseen dangers.

To succeed in these hard times when economic concerns are paramount, any new program must be productive; it must be able not only to pay off its initial investment but to generate new wealth. As we now see it, the first payoff from the development of space communities will be a supply of low-cost electrical energy here on Earth. We should examine the scale of investment which is customary in the electric utility industry, then estimate the cost of seeding a space-manufacturing program, and see whether those figures are in balance.

In 1975 the total installed generator capacity of the United States was about five hundred Gw (five hundred thousand megawatts).[1] When in 1974 the first mild energy shortage struck, a number of studies were initiated to provide estimates of how that capacity would have to grow during the next quarter-century. It was assumed in most of these studies that conservation and rising energy prices would limit energy-growth rates to less than 7 percent per year, which was commonplace in the 1960s.

A working group formed by the Institute of Electrical and Electronic Engineers summarized in a report twelve of these forecasts. According to its conclusions, the installed generating capacity of this country must quadruple, to about 2,000 Gw, during the last quarter of this century.[2] That increase is equivalent to an average growth of just over 5 percent per year.

In order to meet that demand for generator capacity, the electrical utilities of this country will have to spend in this quarter-century about $800 billion, at a rate of $530 per kilowatt.[3] The latter figure is appropriate to coal-fired generating plants; nuclear plants cost considerably more. Eight hundred billion dollars is nearly as much as this country makes in a single year (its gross national product). It's almost twenty times as much as the cost of the entire Apollo Project over the decade during which that enterprise was carried out.

If indeed the establishment of a manufacturing facility in space, able to process lunar surface raw materials, can satisfy our electrical-energy demands, what investment will be required to set up such a facility? By now there have been a number of independent estimates of that investment. One, progressively updated and revised, has been made by the NASA Marshall Spaceflight Center. Another, using NASA figures for launch costs but otherwise independent of the space agency, was made by a study group working in a joint program of NASA, the

129

American Society for Engineering Education and Stanford University.[4] A year later a study group working purely under NASA sponsorship went through a still more detailed estimate of construction time and cost.

All of these estimates were based on a fairly direct approach, not yet taking full advantage of the possibilities for cost-saving inherent in the idea of manufacturing in space. They agreed fairly closely, though, and centered on about $100 billion—only a small fraction of the investment that the utility industry will have to make to satisfy our electrical-energy demands.

These various estimates of investment needed agreed because no great advances in technology were assumed by any of the estimators. Once the total tonnage to be lifted into space was established, the total investment could be estimated from known launch costs and from experience on the development and administrative expenses during the first decades of our space program.

The cost, it seems, would be about twice that of the Apollo project—and we haven't yet talked about ways of further cost-cutting. Although in retrospect Apollo appears as a vitally necessary prospecting expedition, essential to any serious proposal to use lunar resources, it appears that the establishment of space manufacturing would give a much greater payoff: a productive factory in space, with a self-supporting work force of 10,000 people, in contrast to a brief series of daring scientific forays by less than a dozen men. The reasons for that greater payoff are post-Apollo advances in vehicle systems, and above all the "bootstrap process"—using the material and energy resources of space to build manufacturing capacity.

We can see at once that if the materials for Island One must be brought up from Earth, there is no possibility of constructing Island One at an affordable investment cost. An Apollo rocket, costing several hundred million dollars and wholly discarded after one use, could lift payloads into orbit, but only at a cost of thousands of dollars for every kilogram. To go as far as L5 with such machinery would have cost several times more, and to haul freight to the Moon in the days of Apollo ran the cost of every kilogram as high as the price tag on an expensive sports car, some $20,000.

Even if we were to be so optimistic as to suppose that with the investment of many years of time and many billions of dollars we could develop launch vehicles able to operate at a hundredth of the cost of Apollo-vintage rockets, we still couldn't afford to haul up the pieces of a space-colony from the earth. For the shielding alone, the lift costs would be a healthy slice of our gross national product. Clearly, then, to construct space manufacturing facilities mainly out of materials from the Earth would be absurd.

Shuttle places Long Duration
Exposure Facility, seventy
experiments to be left in Earth
orbit for a year.

131

The Skylab Project of the early 1970s yielded a great deal of scientific and technical information, and considerably advanced our understanding of the effects of long-term weightlessness on humans. Its basic rocketry, though, was that of Apollo, so it did nothing to advance the art of launching heavy payloads at lower cost. NASA is now devoting most of its development effort to a project which will push chemical rocketry to a high state of sophistication: that is the Space-Shuttle program. The shuttle is a winged, orbital vehicle intended mainly for scientific missions in low Earth-orbit. It is designed for reuse, at least in part, and it will be particularly suitable for missions in which scientific instruments of large size must be recovered from orbit and returned safely to Earth. In the course of developing the shuttle, NASA is putting a great effort and several billions of dollars into the design, testing, and perfection of what are called "SSME's": space-shuttle main engines. These are not very large engines, in comparison with those of the Saturn 5's which launched the Apollo flights, but they are a great deal more efficient. They operate at an internal pressure as high as modern materials can stand, and at temperatures close to the material limits. It will be some time before chemical rocketry pushes much beyond the performance figures that the SSME's can attain.

The shuttle is designed as a two-stage vehicle, and its first stage is a pair of solid-fuel rockets which after burnout are to be soft-landed by parachutes in the ocean and then (with a probability which only experience can tell us) are to be recovered and reused.

For some time now NASA has been studying designs for a freight vehicle based on shuttle engines: a "shuttle-derived heavy-lift vehicle" or HLV in the language of the rocketeers. It would be a booster, not necessarily manned, which could lift about a hundred metric tons to low-Earth-orbit. The HLV would not be a large vehicle; its height on the pad would be half or less than that of an Apollo-Saturn 5. It would be capable of higher performance than Apollo, its first stage possibly consisting of shuttle solid rocket motors and its second powered by SSME's. There are alternatives, also, to the solid rockets.

Launch of shuttle-derived
heavy-lift vehicle (HLV). Large
external tank carries all
fuel for main engines.

Within the present stage of the art the first-stage engines could be liquid fueled, burning kerosene or ammonia and liquid oxygen. Especially in the latter case, the first stage would release fewer pollutants to the atmosphere, and its fuel would cost far less than that of the solid rocket motors. Either way, the HLV could be built on a rather short time scale, taking advantage of the great effort which has already gone into the development of the SSME's.

NASA is presently advertising a cost of about 20 million dollars for a shuttle launch, assuming complete recovery and reuse of all the hardware required. The SSME's would cost several million dollars each, so for economy they should be recovered from orbit. The latest HLV designs show the SSME's mounted on a re-entry shield, so that after lifting freight to orbit the engines could be recovered by atmospheric braking followed by pop-open parachutes, just as the Apollo command modules were safely recovered with the returning astronauts inside.

In May 1975, at a Princeton University Conference on Manufacturing Facilities in Space two professional rocket designers with many years of experience at NASA presented their estimates for the kind of vehicle needed both to reach low orbit and to go beyond it to L5 or to the lunar surface. Hubert Davis, from the Johnson Space Center in Houston, presented data from several NASA and industry studies on HLV conceptual designs.[5] A. O. Tischler,[6] now retired after many years of service at NASA, discussed a

Rocket-powered lander delivering equipment to lunar surface.

chemically propelled "tug," an engine and control system small enough to be placed in orbit by the HLV and then capable of moving payloads of various shapes and masses from low orbit to L5. To go from lunar orbit to the lunar surface we will also need a "lander," another small vehicle quite similar to the tug. The early estimates on the investment needed for Island One and its early successors were based on just those few vehicles: the space-shuttle, which made its first free flight in 1977, the shuttle-derived HLV, the tug, and the lander, the last two being small chemical-rocket vehicles well within the present range of engineering knowledge,

At the 1975 Princeton Conference it was confirmed that the cost of putting a ton of payload on the lunar surface would be about twice as much for the same load placed at L5, and that the cost to locate at L5 would be about the same as to place a payload in geosynchronous orbit, above a fixed point on the surface of the earth. In later, more detailed NASA-supported studies in 1976 and 1977, these estimates were checked further. Remarkably, it has been found with each successive study, as the engineering has become more complete and the cost-estimation more professional, that the cost estimates for the establishment of Island One have come down.

The most recent work traced a program in which Island One would be preceded by smaller habitats, down to the size of a small space station. These habitats, the first

transportable by the space-shuttle, would be temporary quarters for a workforce whose first priority would be to set up manufacturing in space, so that the program could begin to return profits and quickly pay off the investment made in it. Only after the program was solidly established on a paying basis would the productivity available in space be diverted even in part to the construction of something as luxurious as Island One. In that scenario, it might be one or two decades after the initiation of space manufacturing before Island One and its counterparts would be completed. Apparently, by adopting such an approach the investment required to reach the "ignition point," after which the profits from space manufacturing would sustain further growth, would be cut to only a small fraction of the amount necessary for the construction of Island One as an initial project.

By now we see clearly, I believe, the logical building-blocks in our program of space manufacturing. We can put them together in different ways, and in order to get the greatest payoff for the least investment we'll be studying all the possibilities right up to the moment when the final planning decisions have to be made. Let's look at those building-blocks one by one, though, because they're likely to turn up in any final program plan.

At the 1975 Princeton Conference and at the Summer Study of the same year "refueling" calculations were made. These indicated that when liquid oxygen derived from lunar materials is available at L5, both the cost and the number of launches required from the Earth can be reduced greatly. In fact, when oxygen from the industrial activities at L5 does become available, it will so greatly reduce the cost of tug operations that the chemical tug will perform at a level otherwise unobtainable except from an advanced nuclear rocket. This fact may dictate that the first industry processing lunar materials extract the oxygen alone. The potential savings from that method have not been put into the cost calculations made so far.

The idea of using lunar oxygen for chemical rockets isn't new, by the way. Robert Goddard thought of it a half-century ago, and Arthur Clarke brought up the same idea some years later.

When we look into the economics of space manufacturing, we find that over a few years several million tons of lunar material must be processed. To keep the investment cost down and to keep the number of shuttle and HLV flights within NASA's "traffic model," though, we'd like to hold the lunar installation to not more than a few thousand tons.

The installation on the Moon must therefore be able to launch during a few years a thousand times its own weight. No rocket within present technology could achieve such a figure. We must design instead a transport device that can launch payloads from the Moon without itself ever leaving the surface.

Before we go into the details of the transporter, we should consider how the "bootstrap" principle of establishing a launcher on the Moon can yield a growth of space habitats and of their products without further drain on the resources of Earth. Clearly the first such launcher must be built on Earth, tested and perfected here, and then launched to the Moon and reassembled there. By its presence it will then permit the construction of the first space manufacturing facility at an affordable cost. Once the first habitat is in place at L5, one of its first products, logically, will be additional transporter devices. The cost of moving them from L5 to the Moon will be substantially less than that of bringing more transporters from Earth, and as the total installed cost will be dominated by that of transportation, Island One will become the favored location for their production.

In order to rid ourselves of what Isaac Asimov calls our "planetary chauvinism," we should consider why the Moon, though it is necessary as a materials source, is less suitable than L5 as a site for industry and human habitation. We can be rather quantitative about some of the reasons.

First, the cost of transporting workers and their families to the Moon, and the cost of transporting from Earth the necessary machines and tools, liquid hydrogen, chemical processing plants, and an initial construction station large enough to build a habitat, would all be roughly twice as high as for transport from Earth to L5, so the amortization cost of all such equipment and materials would be far higher on the Moon than at L5. In turn that would increase the price of any products of lunar industry.

Second, any objects which the Moon could build would then have to be lifted off by rocket power. That would limit them to comparatively small sizes; in contrast, the L5 communities could build objects of mass up to tens of thousands of tons, could assemble and test them in their final form, and could then move them to any free-space location where they would be used. Lift costs by rocket from the Moon would be many times higher than the transport costs by mass-driver of the corresponding raw materials.

Third, all the construction efficiencies at L5 which I have described depend on the availability there of constant, dependable solar power, for all energy needs. On the Moon, solar power would be turned off for two weeks out of every four. Though ultimately it will be possible to obtain electric power at any point on the Moon from power lines drawing from solar stations on the "day" side, electric power on the Moon will necessarily be more expensive than at L5, because on the Moon one will have to build two or three solar stations to obtain constant electric power, instead of just one. The problem of supplying the equivalent of sunshine for agriculture, and heat for chemical processing, during the lunar night, will increase further the costs of operations on the Moon.

Gravity on the Moon is a problem for several reasons. It cannot be turned off, so all the possibilities of containerless processing, the building of large fragile structures, high-purity zone melting, and the other attractions of zero-gravity are forever denied to lunar industry.

The inescapable lunar gravity poses a further problem for any large work force located there: it is too small to keep muscles and bone in good condition without strict exercise, and yet it is enough to prevent easily obtaining one gravity by rotation. In free space, for a habitat of modest size, the cost of rotation to imitate Earth's gravity would be only a small addition to the cost of enclosing an atmosphere. Yet on the Moon to accomplish the same result we would have to build a relatively heavy structure supported on massive bearings.

When we consider that any lunar employee will have to put up with no sunshine, or artificial sunshine, for two weeks out of every four, that his transportation cost to the lunar surface will be about twice as much from the Earth, and that he will probably have to spend a considerable amount of time in hard exercise to avoid losing muscle tone, we can see that industry on the Moon will have a difficult time competing with industry at L5. It will have advantages only for such specialized products as mass-drivers and their solar power plants. The Moon seems, therefore, likely to remain an "outpost in space," similar in some respects to Antarctic scientific colonies.

In the long run, as the communities continue to grow in numbers and size, presumably the lunar station will grow also. For nearly all products it will be unable to compete economically with the L5 facilities, because of its permanent disadvantages of intermittent solar power, confinement to a nonzero-gravity for construction, and greater remoteness in terms of rocket transport. It will have a great advantage for just one class of products: those whose end use will be on the Moon. Probably the first of these products will be the transporters, and the second may well be solar-power plants for local use. In the long run, it seems logical to assume that solar-power stations will be located at several points around the lunar circumference, linked by transmission cables, to provide solar-electric energy without interruption. There may also be a possibility of locating stations on a high peak near one of the lunar poles, where sunlight would be available more nearly full time. All such possibilities are, though, for a later period; at first the lunar operation will presumably be confined to a single location, from which the miners and engineers will never stray very far.

As the economic picture grows, the reader will see that the success or failure of the entire space-manufacturing concept rests on the bootstrap-principle, and, therefore, on

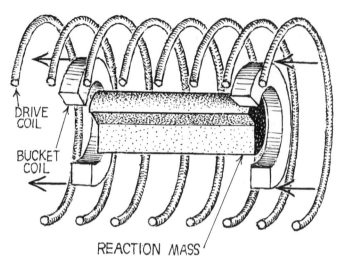

DRIVE
COIL

BUCKET
COIL

REACTION MASS

Mass-driver: Current in drive
coils makes magnetic field that
pushes on currents in bucket
coils, giving acceleration.

the transport device which must transfer lunar materials to
the processing plant and industrial site at L5.

For convenience I call this device a "mass-driver." As
presently conceived[7] it is a kind of recirculating conveyor
belt. By the action of magnetic impulses driven by electric
energy, it can accelerate a small "bucket," containing a
payload of compacted lunar material, to the lunar escape
speed of 2.4 kilometers per second. Then, after final
guidance and precise correction of errors in direction and
speed, the bucket will release the payload, slow down to a
relatively low speed, and be returned to pick up another
payload. The key feature in such a method is that nothing
expensive will ever be thrown away. A bucket can be
extremely costly, and yet will contribute little to the costs
of launching. As the numbers work out, each small bucket
will be re-used every couple of minutes. Even if each one
were to cost as much as a million dollars, that cost
amortized over a few years would add only pennies per
kilogram to the cost of launching lunar material into space.

The mass-driver is a device which could well have been imagined a century ago, as soon as physicists had achieved a good understanding of electromagnetic fields. An early variant of it is described in a publication fully twenty-five years old, by that dean of science-fiction writers (and at that time active working scientist) Arthur C. Clarke.[8] In the *Journal of the British Interplanetary Society* Clarke worked out the basic mechanics of electromagnetic launch from the Moon, and compared the problem to military research then in progress on electromagnetic launching of aircraft from carriers.

Three developments have brought the mass-driver from the realm of science fiction to that of possible practicality. The first is the notion of recirculating buckets: that could have been worked out at any time, and I am still searching for evidence that someone may have written it down many years ago in some publication not yet known to me. The second is the development, just within the past decade, of superconducting wire in commercial quantities. Only now is it possible to build a magnet out of superconductor, and have that magnet operate continuously with a high magnetic field in the absence of a power supply. For the buckets, the superconducting coil will constitute a "handle for the bucket," because it will set up a constant current which external pulsed magnetic fields can grab.

The third necessary development is a curious one: it would have been possible many years ago to accelerate an object by magnetic fields, but for the lunar-launch problem the difficulty was how to guide it. At the necessary high speeds, wheels would fly apart; frictional contact would waste too much energy and generate unwanted heat. The solution lies in an idea first published by a French engineer, Emil Bachelet, more than sixty years ago. That concept, "dynamic magnetic levitation," consists in the observation that if a permanent magnet moves rapidly near a conducting guideway (which can itself be a simple, curved aluminum trough) its magnetic fields generate induced currents within the guideway.[9] Those currents in turn produce magnetic fields, which act to repel the magnet and so produce a lifting force. The higher the speed, the more the lift and the lower the drag. Within the

Lunar mine and mass-driver.
Crew habitat and machinery
tunnels covered by lunar
soil for shielding.

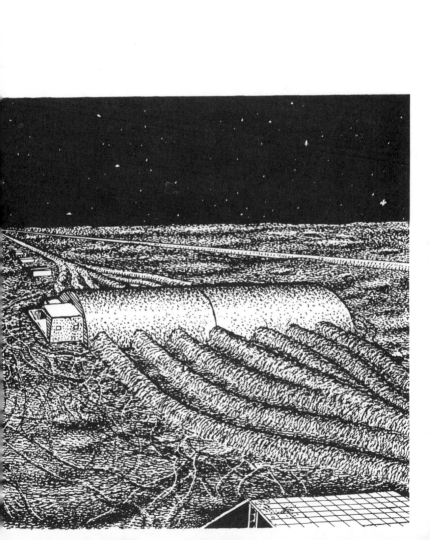

past few years design studies of this concept have reached a fairly large scale, with model "magneplane" guideway systems in operation in several countries. The magneplane or "electromagnetic flight" concept has arrived at just the right time for use in the mass-driver.

If we follow the construction of the mass-driver, we may see it in spectacular operation under test on the Earth. It will be a slim, lightweight tube surrounded by coils, no bigger around than a dinner plate, but many kilometers in length. At intervals there will be small capacitors for the storage of electrical energy, and every coil will be connected to a transistor-like solid-state device to pulse that energy into the coil as the bucket goes by.

We may view the mass-driver only through a window, because it will be designed to operate in the near-perfect vacuum of the lunar surface, and here on Earth can receive its final tests only in a vacuum chamber. Near the "injection" end a bucket will slow to a halt for a fraction of a second; a mechanical conveyor belt will remove it from the guideway for checking, automated inspection, reloading with another payload, and balancing. In its place the conveyor will set another, preloaded bucket. The first accelerating coil will pulse, and then as the bucket passes through each successive coil it will interrupt a light-beam, to trigger that coil and push the bucket to a slightly higher speed. The interrupted light-beam principle is the same one that's been used for many years to hold open doors as people enter elevators. When the bucket reaches full speed it will slow a little to release the payload, then will be deflected, will be braked rapidly by decelerating coils, and when slow enough will go round a gentle curve and be returned at moderate speed to the starting point. Its payload and those that follow will subject the "catcher" to a steady battering that will average to a force of more than four tons.

To supply the lunar mass-driver with electricity, the alternatives are a solar-cell array or a small nuclear plant. We won't need a great deal of power—only about a tenth as much as a typical generator in a power-plant on Earth. The latest studies indicate that a solar-cell array will be so much lighter than a nuclear plant that solar power will be preferable, even though it can only operate during the lunar daytime. As far as can be seen at present, that is the only place in the entire space-manufacturing concept where nuclear power can even come close to being cost-effective.

Like its cousins, the particle linear accelerators used in high-energy physics laboratories on Earth, the mass-driver can still work even if some of its coils fail to operate. We plan to add extra coils along the length of the machine, and in normal operation those coils will be turned off, sitting quietly as spares. In case of a component failure one or more of the spares can be switched on, so that the mass-driver can continue to operate with high reliability. In the maintenance period, probably during the lunar night, the repair crew will go over the machine and replace anything that has gone bad.

In the whole space-manufacturing concept, everything except the mass-driver is a variation of something we've done before. The rockets are conventional, and the manufacturing operations are novel because of their location in space, but are otherwise at least analogous to bridge-building and other operations on Earth. The space-habitats are unique in shape because of their use in vacuum and in zero gravity, but otherwise have their analogs in ship-building, aircraft, and the construction industry. No one's ever built a mass-driver, though, and because of that we have to work through all the basic theory of that machine, and make working models at each stage of development in order to be sure that our thinking isn't going astray.

After I published an article including the mass-driver concept, in 1974, little was done to explore it more thoroughly until 1976, when I led a NASA-supported study investigating possible "show-stoppers" in the space-manufacturing concept. In that study I had the great good fortune to work with Dr. Henry Kolm, of M. I. T., and Dr. Frank Chilton, of Science Applications in California. Kolm and Chilton had been the leaders of groups applying the ideas of magnetic flight and linear electric motors to new possibilities for high-speed ground transit systems. Their groups had developed successful working models, as well as a great deal of basic theory organized in published reports and articles. It's a sad commentary on the decline of an American sense of drive and courage that both projects were killed by the government's Office of Management and the Budget during the early 1970s. At that time leadership passed to Japan and Germany, and with over a hundred million dollars being spent every year by each of those countries on magnetic-flight research, by 1977 there were full-scale magnetically flown test vehicles in operation in both. If belatedly we decide we need magnetic flight to solve our rapid-transit problems in the United States, we'll then have to spend our dollars abroad, with unfortunate effects on our balance of payments, to buy back the developed technology that we could have had for ourselves if we'd been wiser.

With the professional experience and expertise of Kolm and Chilton brought to bear on the problem, in 1976 we were able to answer the most important single question about mass-drivers: Was the idea fundamentally sound and practical? Both experts were quite sure the answer was yes. Kolm suggested that we switch to an "axial" geometry for mass-drivers; in the axial case all the coils would be circular, and the drive forces could be higher. Both men were sure that my old calculations on bucket-acceleration were far too conservative; in their estimation we could achieve accelerations of several hundred gravities, shortening the length of the mass-driver accelerators.

In late 1976 and early 1977 I was able to devote a great deal of time to mass-driver research, under the best possible circumstances. I was on sabbatical leave from Princeton, and had accepted an invitation kindly extended by M. I. T. to become the Hunsaker Professor of Aerospace for that year. It was a great opportunity for close cooperation with Henry Kolm, and we worked together throughout the year, our locations at M. I. T. being only a block apart.

My main effort was on mass-driver theory. In completing the articles from the 1976 NASA study I worked out the optimization of masses, and so learned how best to design a mass-driver in order to get the highest performance with the lowest possible weight.

During the first semester of 1977, at the invitation of

Professor Rene Miller, Chairman of the Aerospace Department at M. I. T. and President of the American Institute of Aeronautics and Astronautics, I gave a series of seminars exploring the questions of acceleration, guidance, design, and applications. Those seminars formed the basis of a 1977 NASA-supported summer-study task-group effort, in which Henry Kolm, Stewart Bowen, several excellent students, and I worked together to put the seminar-results in the form of usable computer programs, and to further extend our knowledge as far as possible.

Meanwhile, we reached an exciting new stage in understanding: the building of the first working model. In the winter of 1976–77 Kolm and I designed an axial mass-driver about as long as a cross-country ski. The comparison is appropriate for another reason: that winter turned out to be the hardest in living memory, and my recollection of it is of great quantities of snow and ice. We had no construction budget until months later, so in January 1977 we enlisted the unpaid volunteer help of a number of students,[10] and of a young post-doc, Bill Wheaton. Our materials were from the scrap-pile in Kolm's laboratory, supplemented by odds and ends like copper plumbing pipe, the brushes from an automobile starter motor, and capacitors of the kind photographers use in their flash-guns.

By early May the model was done, and we demonstrated it at the last of the seminars. Then it was brought to Princeton, and for the next several months traveled quite a bit. At Princeton it became the star performer at a large conference, and was photographed in action by several television crews. Then it was shipped to California, and climaxed the final briefing of a 1977 NASA study on space-manufacturing, at the Ames Research Center. From there it traveled to Los Angeles, and performed (flawlessly) before an audience of a thousand people, invited by Governor Jerry Brown to a celebration of "California in the Aerospace Age," on the day before the first free flight of the space-shuttle orbiter.

In the model, the bucket accelerated from zero to eighty miles an hour in a tenth of a second. Significantly, the acceleration in that first model was already higher than my

High-acceleration working
model of mass-driver.

estimates of several years earlier for an "ultimate" lunar
launcher. Through all these travels two students, Kevin
Fine and Bill Snow, carried out the setup and operation
tasks. Later in 1977 Kevin continued the work and completed a master's thesis on the subject of mass-drivers.

By then, a modest amount of NASA support for research
and development of mass-drivers was available, and we
began a joint program at Princeton and M. I. T. to build a
high-acceleration model. By the beginning of 1977 I felt
confident enough in our understanding of mass-drivers,
and the calculated performance figures had improved so
much, that I could apply the concept not only to the
lunar-materials launcher but elsewhere in an updated
version of a lowest-cost, maximum-payback plan for space
manufacturing; more on that a little later. Now let's trace
the flow of material from the Moon to and through a
processing facility in space.

Lunar mining need not be a large-scale operation. Chemical processing can be done at L5, and all industrial slag
produced there will be usable as a matrix for crop growth,
as shielding against cosmic rays, or as reaction mass for
mass-driver engines in free space. For that reason there
will be no need for initial separation of the lunar surface

material by high-temperature processing. Experts in commercial ore-processing who studied the problem believe that it will pay to "beneficiate" the lunar material, carrying out separation by sieving or magnetic effects, to increase the fraction of useful elements. After those basic operations the material can be compacted, bagged, and prepared for shipment.

Dr. David Criswell has studied the problem of containing the lunar material during its travel from the Moon into space, and has worked out the details of a facility on the Moon that would make glass-fibers and weave them into bags for the material. Fortunately, typical lunar sites have large quantities of glass lying about, in the form of sand that can be melted by solar furnaces.

When one first hears the phrase "mining the moon" one thinks in terms of vast open pits, scores of giant machines, and a scale of operations comparable to our great terrestrial mines. The reality will be far more modest. If the surface is excavated even to the depth of a shallow gravel-pit, and a million tons or more are removed every year, in several years of operation the whole operation will still be so small that you could walk the length of it in a few minutes. Mining experts who have looked at the problem consider the lunar mine so small-scale that it will hardly keep one bulldozer occupied.

As long as we demand great quantities of elements which are rare on the Moon, there will be no need for detailed assaying at the lunar surface. The *average* lunar soil (for example, the so-called "fines" brought back by *Apollo 12*) is about one-third metals by weight, and almost a fifth silicon, useful for making solar cells to convert sunlight to electricity.[11] Oxygen is the most plentiful element on the lunar surface, and so will be an abundant and very useful "waste product" of the processing industry in space.

Television and personal report have shown us that men can work in space suits only slowly and inefficiently; if the lunar outpost is to carry out its tasks quickly and effectively we must so plan the activity that space-suit operations are reduced to a minimum. The most time-consuming task may be the assembly and checkout of the mass-driver. Within a modest mass-budget, a circular cylinder of aluminum large enough in diameter to serve as an assembly bay could be delivered to the Moon in sections, among the early payloads. In such a cylindrical tunnel, covered over with lunar soil for cosmic-ray protection, the mass-driver could be assembled and electrically tested.

By the time the cooks, the doctors, the communications experts and the other necessary service personnel are added, the lunar work force during the construction phase may total about fifty people. After construction is finished and the lunar outpost settles down to steady operation, the best estimates are that eight or ten people will be enough. On a typical work-shift there may be one person monitoring the automated operation of the mass-driver, while another controls a mining vehicle by television and radio. The two may be in the same room, at control consoles, and while the work goes on may be swapping stories and passing the coffee-pot back and forth.

Path of lunar material to
second Lagrange point.

In most respects the lunar base will be the most remote and difficult to get to of all the locations where people will be working. It's unlikely to become a backwater, though. Scientists will visit the Moon both to do basic research and to carry out assays and surveying. Construction crews will visit each time the mass-driver gets upgraded. As we now see it, the mass-driver first located on the lunar surface will be capable of moving over a million tons of lunar material each year. Its power supply will be a lot heavier than the machine itself, though, so it makes sense to give it only a fraction of its final power initially, and add solar-cell arrays as the industry in space expands.

When installed and operating on the Moon, the mass-driver will launch its payloads at a slight downward angle. Their speed will be so great that they'll rise rather than fall, and after a free flight of a minute or so will be many kilometers away. There they will pass through a correction-station, where their positions will be measured very accurately, and their speeds and angles will be corrected by the same electrostatic methods that are used in steering the beam of electrons in a television tube. The latest calculations show that after such steering the payloads will be able to hit a particular point in space with an error of only a few meters.

Climbing out against the pull of the Moon's gravity, the payloads will finally escape from it to free space at a relatively low speed. What's the best point to aim for? The best target seems to be the second Lagrange point, L2, out beyond the lunar farside. There a collector will be maintained in position, maneuvering to follow the slowly changing stream of lunar payloads as the trajectory changes over the period of a lunar month. When several thousand tons of material have accumulated at L2, they'll be ferried to L5 by a low-thrust tug; that tug may itself be driven by a small version of the lunar mass-driver.

Newton's laws tell us that a machine which can accelerate and launch material with a high velocity can be used as a reaction engine, like any rocket. The mass-driver, with its tons of steady force, will be quite effective for conveying large payloads in free space. Its performance, measured in terms of exhaust velocity, will be about that of the space-shuttle's solid rocket motors.

The lunar machine isn't designed to be a rocket engine, and in the course of the intensive theoretical work on mass-drivers I became interested in calculating the performance of a mass-driver tailor-made to supply thrust, as a tugboat-engine in space, driven by solar power. The numbers looked so attractive that in 1977 I included them in an article that is typical of our modern ways of approaching the space-manufacturing problem.[12]

Let's be realistic about our plans of realizing the humanization of space. First of all, no one's likely to subsidize the construction of space habitats for their own

Stream of payloads
leaving lunar mass-driver.

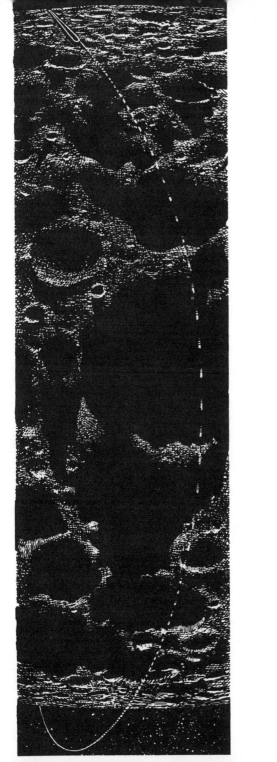

Mass receiver at Lagrange
point L2, beyond lunar farside.

sake no matter how attractive they may be. If they're built, it will be for the same reason that most new housing is built on Earth: there's an industry, or several industries, that need workers, and so a market exists for housing for the workers and their families.

If there is a need for products, in large tonnages, that will find their use in high orbit or beyond, we should search for the most efficient ways to set up the manufacturing and transport systems to build and locate those products. How can we minimize the necessary investment? By using, as far as possible, the one vehicle system that is already under development: the space-shuttle. During the decade of the shuttle's development, it has been planned for a traffic model ranging from 60 to 120 flights per year. If a particular orbiter must spend a long period in orbit, to carry out planned experiments, it can only be used in a smaller number of flights per year. To accommodate not only NASA's present (much reduced) schedule of launches, but also a program of space manufacturing, some additions to the presently planned fleet of five orbiters may be needed. When the shuttle was first planned, it was thought of as a vehicle to lift components of a space station. More recently, as budgets have declined and the space station has shrunk to something more like a workbench in space, the shuttle traffic planning has been changed. Now the shuttle orbiters are thought of as doing double service, bringing experiments into space and remaining with them as temporary space stations. In terms of traffic efficiency, that is a bit like traveling to Europe in a 747 on a week's vacation, and then keeping the airplane on the ground the whole time in order to use it as a hotel. NASA has no choice, under present budget limitations, but if orbiters could be used literally as shuttles, bringing equipment into orbit and then returning as quickly as possible, a fleet of three or four more would be enough to double the number of flights, beyond the presently planned sixty a year.

In the article "The Low (Profile) Road to Space Manufacturing," I outlined a way to attain a high level of production in space over a period of a few years, within a traffic model of about sixty flights per year of shuttles.[12] In the later years of that plan many of those flights would be of the shuttle-derived HLV, the shuttle being retained mainly as a transport for people. Its cargo bay is about the same size and shape as that of a DC-9 aircraft, and if filled by a passenger compartment could carry—for the short flight into orbit—about the passenger load of such an aircraft.

The "Low Profile" article built heavily on the results of a 1976 NASA study on space manufacturing. There for the first time we obtained the numerical data on the sizes and weights of processing plants in space, and on the number of people necessary for a manufacturing program with a certain tonnage of output per year. In 1977, in a much larger study, a group working under the direction of John Shettler of the General Motors Corporation followed up the "Low Profile" article by a much more detailed investigation, calculating the equipment payload and the passenger list for each flight. These are early steps in what is likely to become a continuing effort, as we search for the most cost-effective ways of realizing space manufacturing. For that reason it doesn't make sense to list a great many of the published numerical results. Instead, I'll continue with the building-blocks we now think of using.

All of the equipment we must locate on the lunar surface must first be hauled to lunar orbit, together with the rocket fuel needed to soft-land the equipment. The shuttle can't do that job, and if we were to use a rocket-powered tug the shuttle would have to lift all the fuel for the rocket. We plan a substantial cost-saving by using a small mass-driver, of very high performance, to carry out that interorbital transfer. The mass-driver would be carried to low Earth orbit in several shuttle payloads, and would be assembled in orbit, from then on to ferry equipment out to the vicinity of the Moon.

Where to find, though, the reaction mass for the mass-driver to throw out? It has to throw something, in order to develop thrust. The answer seems to be to use something that would otherwise be thrown away: the shuttle external tanks. The orbiter vehicle has engines (the SSME's) but no fuel tanks for them. When it rides into orbit it does so on the back of a much larger object, a big canister shaped like a fourth-of-July rocket. That canister contains hydrogen

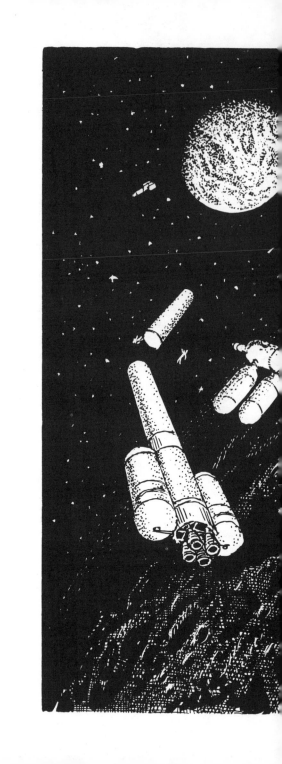

Workshop in lunar
orbit. Power supplied
by solar-cell arrays.

and oxygen tanks from which the SSME's draw their fuel, and when the shuttle is almost at orbital height and speed that fuel is exhausted. The final tiny push into orbit is done by much smaller steering rockets carried by the orbiter, and at the moment of burnout the external tank suddenly becomes surplus, after a brief but glorious life of less than twenty minutes. It happens that the empty weight of the tank is actually greater than the whole shuttle payload, and it seems a shame to let that weight go to waste.

In the "Low Profile" plan, the tanks would be carried into orbit, at a very small cost in shuttle payload. We would set up a storehouse of empty tanks in orbit; some would be fitted out as living quarters, each tank providing about twenty comfortable, private apartments for as many workers. In Shettler's plan, those modular apartment-houses would turn up everywhere in the early days of

Moment of separation, as
solid-rocket boosters drop from
shuttle-derived freight vehicle to
be recovered by parachute.

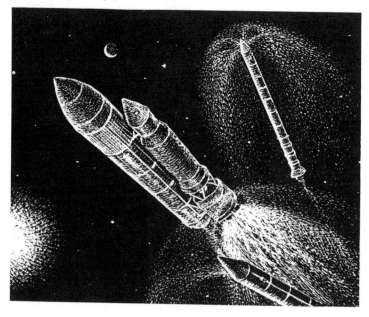

space manufacturing: in low orbit, for the training and final screening of workers in the special world of zero gravity; in high orbit, for the workforce tending the processing plants; at L2, for times when the mass-receiver there may need repair; and on the lunar surface. As soon as lunar material became available in space, it would be used to shield the apartment-modules from cosmic rays, and before that there would be minimal shielding, enough to protect against solar flares, made up of dehydrated foods stored for later use.

Most of the external tanks would end as reaction mass, in pelletized or powdered form. In a typical (unmanned) ferry operation, several hundred tons of equipment, accumulated from shuttle payloads, would spiral up to lunar orbit, over a time of several months, with the expenditure of a somewhat greater amount of tankage-mass, each small pellet leaving the mass-driver tug engine with a speed much greater than that of a rocket exhaust. Dumping the equipment in lunar orbit, the mass-driver would return in a much shorter time, arriving in low Earth orbit to pick up a load for another round trip.

In our present thinking there would be several stages in the setup of manufacturing in space, and if an insurmountable problem appeared at any stage the program could be terminated there. We don't really expect any such problems to appear, but it is far easier to arrange funding if there are well-defined milestones, tests each of which has to be passed before the final goal is reached.

The first stage is the setup of the lunar mass-driver and the beginnings of the transport of lunar materials into space. That seems to require only about two years' worth of shuttle flights. Once that milestone is passed, we'll be able to bring into high orbit about ten times the amount of material that the shuttle can lift. Already from that point on there'll be plenty of mass for shielding, and plenty of "fuel" for mass-driver reaction engines.

The second stage is the beginning of chemical processing of lunar materials into pure metals, glass, and oxygen. That takes about another year's worth of shuttle flights, to lift the processing equipment, solar power arrays to run it, and other essentials. When that stage is reached the

161

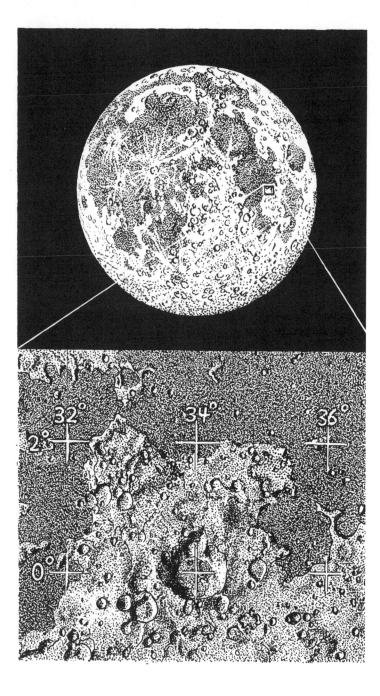

number of workers in space will be something like one or two hundred.

Now comes another application of the "bootstrap" method. The most complicated and sophisticated pieces of equipment needed in space—things like mass-drivers and chemical process plants—turn out to be rather light. It makes sense to build them and test them on the Earth, and lift them to orbit with the shuttle. The heaviest pieces of equipment needed in space seem to be solar-cell arrays, to power both the lunar mass-driver and the processing plants. The first pilot-plant in space will already be turning out each year several thousand tons of metals, silicon, and oxygen. We plan to use all three to bootstrap our way to a much higher level of productivity.

The metals and silicon will go into solar-cell arrays. Those we will use to upgrade the tonnage per year that the lunar mass-driver sends out, and to equip new duplicates of the original space processing-plant. The oxygen will be used in several ways: as the heaviest part of the fuel burned by the rocket tugs and landers; as the heaviest part of the water that will be needed by the work force in space; and as an ideal reaction mass for the increasing traffic of mass-driver tugs hauling freight in space.

It seems that by this kind of cost-saving approach we can build up to a level of processing a million tons or more of lunar material each year, over a period of seven or eight years, without ever exceeding the lift capabilities of the shuttle. How about the economics? In our traffic model we'll be paying about $1 billion in shuttle launch costs each year, over a seven-year period. At the end of that time, though, we'll be producing about a third of a million tons every year of finished products, and relocating them either in geosynchronous orbit or wherever else in nearby space they will be used. A good cautious estimate would be to assign those finished products a value of around a hundred dollars a kilogram; the lift costs alone, to bring a kilogram into high orbit, are in that general range even for very

Possible site for lunar mass-driver, near boundary of mare and highland regions, for greater choice of minerals.

advanced, totally re-usable rocket concepts many times larger and many years later in time than the shuttle. With those numbers, the manufacturing facilities in space will be producing $30 billion every year in value. A bargain indeed.

How soon could it all happen? Both in 1976 and in 1977 the studies independent of NASA, but supported by that agency and closely cooperating with it, worked out program-plans based on slow and fast rates of decision-making. It seems to be agreed generally that there are far greater uncertainties in the political decision-making process than in the technical areas. With rapid decisions, both the '76 and '77 studies agreed that the first liftoff of equipment destined for space manufacturing could occur as early as 1985, and that the first substantial payback in the form of products made in high orbit could occur as soon as 1991. On that time-scale, by the mid-1990s the construction of Island One, as a more comfortable, long-term habitat for manufacturing workers and their families, could be done almost as an aside, with the diversion of only a few percent of the manufacturing productivity that would exist in space at that time. There's no answer to the question "What's the longest it might take," of course, except "Never." The more leisurely program-plans put Island One around the year 2010. To those of us who feel that space manufacturing offers great potential for human benefit, such a delay seems nearly criminal, but on the time-scale of human existence a mere fifteen years is hardly the blink of an eye.

9
FIRST TASKS
FOR
ISLAND ONE

As the first new world in space takes shape, over a period of several years, surely the moment of "sealing" will be planned for and celebrated. Oxygen long stored in liquid form will be allowed to enter the sphere, and pressure throughout the living and agricultural areas will slowly build toward its final value. Many of the construction

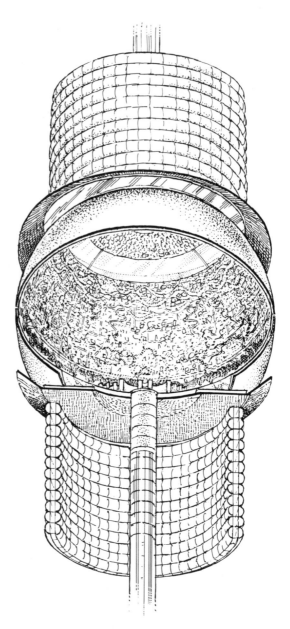

Cutaway view of Island One. Axial
cylinder is air passage and corridor
to docks and industries in zero gravity.

workers may move their activities to the new villages at that time, and enjoy the luxury of roomy surroundings as they complete the apartments and other buildings.

As they work, a small electric motor the size of an automobile engine will apply its power to rotating the habitat, until finally after several months the gravity at the equator will reach Earth-normal. By then, the soft green of growing plants will have turned the valley into something very like a small patch of farmland in springtime.

With the greening of Island One, and the harvesting of its first crops, the long-term residents will come, and for many in the construction work force a time of decision will arrive: a time when the choice must be made, to return to Earth or to stay on to help lead the growth of the new communities in the first permanent human world beyond Earth. Many will choose to return to our planet, to spend and enjoy accumulated earnings; some, though, will probably feel that nothing here can offer them the excitement and the challenge of construction at L5. If human nature and history are guides, some people will make the first choice, will visit for a time on the Earth, and then will be outward bound again to rejoin their friends who may never have left.

Island One, though modest in size, may be an attractive place in which to live and work; certainly there will be few communities whose citizens have so many talents and so much determination. Whatever the attractions of Island One, if it is to take its place as part of the complete human world, it will have to produce, more effectively and efficiently than can be done in any other way, products which are needed urgently by the rest of the family of humankind.

Island One will have a unique economic advantage for just one class of products: those whose end use is in free space or in high orbit above Earth. When we attempt to build such products and launch them from our planet, we must pay heavily in energy. Here on Earth we are the "gravitationally disadvantaged," located as we are at the foot of a gravitational mountain some 4,000 miles high.

For any product whose use is to be in or near free space, high above Earth, production at L5 will save the lift cost, many dollars for every pound produced. A worker at L5, producing at a similar rate to our heavy industry on Earth (more than twenty tons per year) will be turning out a value of several million dollars per year, beyond the intrinsic value of the goods, simply because of the saving in lift costs from the Earth. The "Swiss banker" approach to estimating the value of Island One is the most conservative we can imagine: value the goods produced by taking the lowest possible lift costs for a competing industry which must lift its products from the Earth, and don't assume any extra productivity in space even though we know that zero gravity and automation are almost certain to favor high production. When that is done, and only half the population of Island One is assumed to be engaged in factory work, the products of Island One still come out to be of great value: many billions of dollars each year, enough to pay back the investment in a very few years.

In the very long term, perhaps material products or raw materials from space can be returned usefully and economically to the surface of Earth. That seems to me a rather unlikely prospect—at least for some time—because if L5's industries begin manufacturing material products to be used here on Earth, they will give up their single greatest advantage: their location at the top of the 4,000-mile gravitational mountain at whose feet we stand. Similarly, I see no great advantage in L5 for the zero-gravity processing of very lightweight, high-value products. For those, it makes more sense to lift the raw materials from Earth to low orbit, by way of the shuttle, and then to return them to Earth the same way when they have been formed in zero-gravity. Others have estimated the total market for products of that kind: vaccines, single crystals, and other exotica—and have concluded that even twenty-five years from now that total market will be so small as to require only a few shuttle flights per year for its satisfaction.

Before considering the major industries for L5, we should ask first whether some of the benefits from Island One's construction may appear during the time of its building. There may well be such benefits, and my guess is that they will be mainly scientific. Once the lunar outpost and the L5 construction station are established, with all their facilities for supporting people, for transport and for communications, they will also be locations ideal for other work than that of producing Island One. At their locations scientific research can be carried on at far lower cost than by the exquisitely complex, delicate pieces of "orbital jewelry" which we now have to launch for completely unattended automatic operation. I expect that among the eight or ten people of the mining and transporter-servicing outpost-community on the Moon, at any given time there may well be several geologists and other scientists in long-term residence. Such people could spend half or more of their working time on such practical tasks as bore-sampling the lunar surface, assaying minerals, and planning the optimum locations for materials-gathering. The rest of their time they could spend on pure research as opposed to applied. If our experience on Earth is any guide, these two activities would be separated only by an indistinct boundary, and would reinforce each other, knowledge gained in one area often finding its greatest use in the other. At L5, by the time the workforce has built up to several thousand people, even before Island One is built there might well be fifty or one hundred whose tasks would be wholly or mainly scientific. Some could be maintaining and gathering data from large space telescopes, located just far enough from the station so as not to be occulted by its busy transport craft, but close enough to be reached in a few minutes travel.

Of the aluminum and other metals being produced at the station it would be surprising if some few percent were not allocated to scientific purposes. The first of these might well be the construction of large optical and radio telescopes. I do not think it likely that such scientific efforts, greatly as they would benefit by "tagging along" on the main construction activity, would ever enjoy budgets large enough to pay back a large fraction of Island One's construction cost, but their scientific objectives could be reached at far lower cost because of the existence of Island One's construction station.

For the scientists themselves, the presence of the L5 construction station would certainly constitute a great boon. Typical scientific programs for space research, even

169

those which involve unmanned satellites, cost several tens or hundreds of millions of dollars. In contrast, the "exchange" cost of sending a scientist to L5 could be as little as a few hundred thousand. One gets a "ticket cost" in that range by taking the existing space-shuttle as a passenger carrier, with the published NASA figures for cost per flight, and assuming that the transfer from low orbit to L5 is made by a conventional rocket-powered tug whose heaviest fuel component, liquid oxygen, is obtained as a waste product from the processing of lunar soils at L5.

Years later, when more efficient vehicles are developed, we can expect that the costs for passage from the Earth to L5 will be reduced, ultimately to only a few thousand dollars.

The most recent studies agree that in the early days of the buildup of production capacity at L5 it will be more economical to bring food from the Earth, rather than to attempt to set up agriculture in space. By the time the workforce reaches several thousand people, though, supplying their food from the Earth will begin to strain the capacities of the shuttle-derived HLV at its normal "traffic model" flight frequency. In detailed studies the trade-off between resupply and space agriculture has already been calculated, and it seems fairly certain that by the time of Island One the people who are living in space will be growing most of their own food. Very similar arguments come up when the planners set out the tours of duty for the early construction workers. It seems that we are likely to begin with stay-times in space of a few months to a year, and then will gradually extend to stays of two or three years, family members accompanying each worker. Clearly the balance between exchange time, the degree of luxury of the construction station, and the salaries paid to the construction crew will have to be made with care after considerably more study.

The problems we now face here on the surface of Earth due to the rapid exhaustion of conventional fuels were described in the first chapters. There are natural sources of energy which we do not now fully exploit, and which could be of benefit to us in extending the fuel reserves that now remain. These include geothermal energy, hydroelectric power, the winds, the tides, and solar power. All of these "exotic" sources of energy have serious limitations. Either they are undependable, or the capital cost of using them is too high, or (as is the case particularly with hydroelectric power) their further exploitation could only be accomplished at a very serious ecological and environmental cost.

Two sources of power for the future are now under intensive study: nuclear fission, particularly in the form of liquid-metal fast-breeder reactors and hydrogen fusion, by magnetic containment of a plasma or by laser-implosion of small deuterium-tritium pellets. It would be rash to attempt to guess the probability that one or both of these methods will turn out to be economically viable. Fast-breeder reactors would have a decided environmental impact, and would also affect the political tensions of the world in ways on which we can only speculate. Rather than guess how successful, how acceptable, or how economical one or both of these methods might turn out to be, I will say only that both are high-technology options on which research is now very active. At present, at least $700 million of government money is being spent each year on nuclear energy research, in this country alone.[1] Of that amount, most goes into fission research, the remainder to fusion. One of the difficulties with the breeder-reactor option is that the "doubling time" for converting nonfissionable elements into usable nuclear fuel is estimated to be at least ten or twelve years, while the world need for new energy resources is doubling in a much shorter time. As for nuclear fusion, most responsible scientists working on it hesitate to claim that it might be economical, even if it can be made to work, in less than about thirty-five years. It does not seem to me very likely (and here I express what is necessarily only my own opinion) that either will be able to *reduce* significantly the cost of electric power; the proponents of the two schemes usually argue at most that one day they might be at a par economically with current fossil-fuel plants.[2]

Perhaps surprisingly, it appears that Island One may be in a uniquely favorable situation to provide for us, here on the surface of the Earth, an alternative energy source which might be simpler, cheaper, and more acceptable environmentally than the first two alternatives. The space manufacturing facility could do so by building Satellite Solar Power Stations (SSPS). Satellite Solar Power is a concept that originated in the 1960s, and whose most active champion has been Dr. Peter Glaser, of the Arthur D. Little Company in Cambridge, Massachusetts. [3] The plan consists of locating in geosynchronous orbit, above a fixed point on Earth's surface, a large solar power station. At the station solar-electric power would be converted to microwave energy, which would then be directed in a narrow beam to a fixed antenna on the ground.

At first glance this scheme appears impractical. Without calculation, most engineers would assume that the inefficiencies of conversion, transmission, and reconversion would be so low that no such power station could be economically viable. Curiously, the transmission problem seems to be solvable. Research on high-power microwave transmission has demonstrated experimentally that power can be transmitted at an overall efficiency of at least 55

Electricity from satellite ground antenna
flows to regional grid day and night, in all weather.

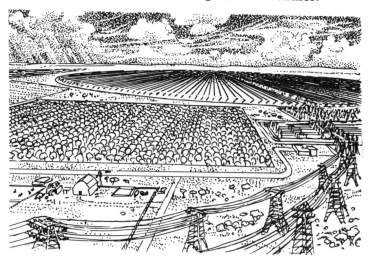

172

percent. [4,5] The target figure for economic viability is not much higher than that, so with moderate development one would expect the target to be attained. The environmental problems of microwave power transmission will have to be studied carefully, but so far they seem to be much less severe than those of radioactive waste generation from fission or fusion nuclear plants. The microwave beam would arrive at Earth with a beam width of about seven kilometers. Its intensity would be modest, less than half that of sunlight. In contrast to sunlight, though, it would be there all the time, even at night or in clouds or rain, and it would be in a form ready for conversion to DC current with a loss of only 10 percent. The antenna region on Earth would be fenced, and outside the fence the intensity of microwave radiation would be no higher than outside a microwave oven with the door closed. One or two kilometers farther away it would be far lower still. Although the beam would be no "death ray," studies would have to be made to be certain that it would have no long-term effect on birds flying through it frequently or nesting in the antenna, and that it would not damage the communications radios of any aircraft straying into it.

Satellite solar power would have significant advantages over its possible competitors, beside the fundamental one of generating no radioactive wastes. Because the conversion of microwave energy to direct current could be done with such high efficiency, only a very small fraction of the total power would be released as waste heat into the biosphere from such an installation. In contrast, generator stations using fossil or nuclear fuels deposit as waste heat in the biosphere about one and a half times as much energy as they put into the power grid.

The market for new power stations during the time when Island One could become productive has been estimated by a number of task groups. For the United States alone, even assuming energy conservation, there will be a need for 65,000 megawatts per year of new generator capacity in the year 1990, and substantially more than that each year a decade later. For scale, the largest single power plant that one normally sees in driving the roads of America is about 1,000 megawatts. The cost of new power plants fueled by coal is roughly half a million dollars per megawatt; nuclear plants are considerably more expensive. Consequently the market for new power plants in the United States alone, assuming prices for coal-fired generators, will be about $33 billion in the year 1990. A satellite solar power station requires no fuel, so its market value may be similar to that of a hydroelectric station of similar size. One of the largest and newest of the hydroelectric installations in the Western world is "Quebec Hydro" at Churchill Falls in Canada. Its price per kilowatt is about three times that of a coal-fired plant, but because it requires no fuel it can supply electricity at a very low rate. On that basis the market for new satellite power stations in the United States at the end of this century turns out to be well over $100 billion per year.

If we include, as we properly should, the additional market represented by the remainder of the industrialized world, and provide for the needs of the nations now struggling to industrialize, the requirement becomes much larger still.

For any power source requiring a large development investment, the potential for long-term growth is important. The SSPS concept appears to fare well on that score also. In the extreme case (certainly not realizable in practice) that SSPS power were to become the sole source of electric energy in the United States in the year 2000, the land area necessary for the SSPS antennas would still be only 0.2 percent of that of the continental United States; that is, about one fifth of the area already devoted to roads. Unlike the roads, SSPS antennas could be located in remote areas where they would not be visually obtrusive. They would be almost fully transparent to sunlight, and would block out microwaves from the land below them, so the areas below them should be usable as protected grazing land.

By contrast, if solar cells at Earth's surface were to be used to supply all our electric power, we would have to cover about forty times as much area, or 8 percent of the continental United States, with opaque solar arrays. The reason is that solar cell electric conversion efficiencies are

174

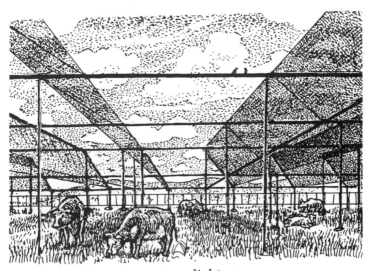

Antenna is transparent to sunlight
and rain, but absorbs microwaves,
leaving fields below safe for cattle.

about 16 percent (instead of 80 percent) and that the average over a year of solar energy intensity in the United States is only an eighth as much as in space.

If satellite solar power is an alternative as attractive as this discussion indicates, the question is, why is it not being supported and pushed in a vigorous way? The answer can be summarized in one phrase: lift costs.

I have discussed the present and the hoped-for figures for lift costs to L5 from the surface of Earth, based on present rocket vehicles and on those which could be developed at low cost with existing engines. Estimates by NASA center on about two hundred dollars per kilogram, for the shuttle-derived HLV. If we don't "go for the Moon" and bring out lunar soils as reaction mass for mass-drivers, we're forced to bring up from the Earth all the fuels needed for the lift from low orbit to geosynchronous. In that case the lift cost to the final location of a satellite power station will be several times higher than to low orbit. (The velocity change needed in order to bring a payload from Earth to geosynchronous orbit is about the same as to L5, so lift costs to either destination will be rather similar also.)

Large power plants could be built in either of two ways: as turbogenerator stations, like present-day generators on Earth, or as arrays of solar cells, converting light directly to electricity.

For use in a power satellite, the most suitable variety of turbogenerator is a "closed-cycle Brayton" system, in which gaseous helium recirculates endlessly between a heater, a turbine, and a radiator.[6] Such systems are rather light and compact as turbines go. Fortunately one such machine has been installed at Oberhausen, in West Germany, and has been working since early 1976.[7] It is heavily instrumented, and will provide plenty of operating experience on which future performance estimates can be based. Studies by the Boeing Aircraft Company, under NASA sponsorship, indicate that a power satellite based on a turbine of the Oberhausen type (that is, right-now technology) would have a mass of about ten tons per megawatt of output power. There is hope, but so far only a hope, that by pushing temperatures higher and using more exotic materials at critical locations that figure can be reduced.

We can assess the state of the silicon-solar-cell art by the fact that such photovoltaic power supplies in operational satellites of the past decade have weighed about ten times as much as an Oberhausen-type turbogenerator.[8] For the Solar Electric Propulsion System space-probe scheduled to fly in the mid-1980s, NASA is hoping to bring the mass of solar cells down close to the Oberhausen-generator figure for tons per megawatt.[9]

If one takes the Oberhausen figure for performance, together with a transmission efficiency of around two-thirds, and lift costs to geosynchronous orbit characteristic of the shuttle-derived HLV and a rocket tug, one finds a transport cost of $13 million per megawatt of installed capacity, lifted to geosynchronous orbit. That is many times larger than for the most expensive power plant now thought of for Earth.

The proponents of satellite power recognize this fact, have represented it accurately in discussions of the topic, and have sought to circumvent the problem by encouraging the vigorous development of lightweight silicon solar cells. Solid-state research moves rapidly, and it may be that ultimately great reductions in the weight of solar cells will be brought about. Even the most optimistic estimate does not suggest, though, that they could be reduced in mass by a factor large enough to make the Earth-launched SSPS concept viable without two more developments: first, while solar cell mass per megawatt of power is being reduced manyfold, their cost must go down by an even bigger factor. In addition to these improvements, lift costs to geosynchronous orbit must come down to a tenth or so of what we could get with the shuttle-derived HLV. To achieve that, it would be necessary to develop advanced space-transport systems for which an investment of several tens of billions of dollars and many years of time would be required.

In giving these figures, it is not my intention to deny the possibility that all these improvement factors could be achieved; I simply do not know. Nor is it my intent to discourage or delay the development of a prototype SSPS; any new technology requires a learning process, and if the basic SSPS concept is to become usable that learning process must go on. Rather, my purpose is to explore an alternative method of the quantity production of economically competitive SSPS units.

Given the existence of Island One, it could produce a satellite solar power station, from lunar surface materials, within the technology limits of the present day; it could simply build large turbogenerators. A complete power station built around a Brayton-cycle turbine would start with solar mirrors, concentrating sunlight onto boiler tubes. Helium brought to a high temperature in these tubes would pass through the turbine, then to a radiator, and be recirculated for another passage. The turbine would drive an electric generator of the conventional sort now found in Earthbound power stations.

If this design is followed, a station built up of several large turbogenerators will be connected to a disc-shaped transmitter antenna. The conversion from low-frequency to microwave power can be made by a large number of small tubes, each like those which power microwave ovens. Operating in the vacuum of space, these tubes will have no need for glass envelopes.

If we take present-day figures for the masses required, a station able to supply 5,000 megawatts to a national power grid on Earth's surface will total some 80,000 tons. It can be assembled and tested as a single unit, in zero-gravity just outside Island One. The work force assembling it will be able to return after each day's work to the comfortable Earthlike surroundings of their habitat.

Studies by the NASA Johnson Space Center, based on projections of technology rather than on the right-now situation, are about twice as optimistic as these figures on mass per megawatt. If they are correct, Island One could turn out about twice the value per year that I've estimated.

Power satellites in twenty-four-
hour orbit stay in constant sunshine
above fixed point on Earth.

The space-manufacturing site will be some distance from geosynchronous orbit. The costs of transport in space, however, are measured not in distance but in velocity interval; in those units even L5, the most distant of the possible sites, is closer to geosynchronous orbit than to the lunar surface. To move so large a mass over the required distance will require a mass-driver, and it could be identical to the one already in use on the Moon. The steady four-ton force produced by that transporter will be quite enough, over a period of months, to move the power station into its position high above a fixed point on Earth. The electric power input to the mass-driver will come from the station itself. The necessary reaction mass to carry out the transfer can be industrial slag, pulverized rock dust, or

178

liquid oxygen, all of which will be available at L5. Return of the mass-driver to L5 for reuse can be made with the help of a small solar power plant. A power plant only about a thousandth the size of the SSPS itself will be quite enough to return the mass-driver to L5 for re-use in a month, so a mass-driver used as a tugboat for SSPS-barges can make several round trips every year.

It was pointed out to me by Mark Hopkins, a young economist from Harvard, that the economics of SSPS construction at L5 requires a fresh viewpoint. In that construction almost no materials or energy from Earth will be needed. Island One, when it is established and operating, will be self-sustaining, and its residents will be paid mainly in goods and services produced at the space community.

The economic input to a combined space community/SSPS program will be the sum of the development and construction costs for Island One, the cost of lifting the material needed from Earth for subsequent communities and for those SSPS components which cannot be made at L5 economically, a payment on Earth to the credit of each person living at L5, representing that portion of salaries convertible to goods and services on the Earth (for subsequent use on trips or, if desired, on retirement) and a carrying charge of interest paid on the outstanding balance in every year of the program.

If Island One and its sister-colonies become the main source for new generator-capacity to supply electricity for the Earth, the question of legal ownership of the SSPS plants ties in to the economics. Geosynchronous orbit is far below L5, and I suspect that any Earth nation using SSPS power will want clear-cut legal ownership of the power-generating facility, once construction is finished. From then on that nation will control the power station and any maintenance operations on it, and will keep the SSPS fixed above a certain point on its own territory, where an antenna is located.

If Island One were to be independent of Earth, it would also be to the economic advantage of the workers in space to sell completed power stations rather than electric power. In that manner they could get a quicker return.

Assembling solar panel arrays and
concentrating mirrors of power satellite.

From the viewpoint of the nation, consortium of nations,
or consortium of utilities which might provide the investment capital to build Island One, it is more cautious
though to assume that the only economic payback will
occur from the sale of power at the transmission lines on
Earth. For many reasons, among them legally binding
treaties which have already been signed by several nations,
it seems wisest to assume that initially Island One will be
tied to the Earth governmentally.

By now the economics of SSPS construction at space manufacturing facilities has been discussed in a technical article,[10] in testimony before Congressman Donald Fuqua's subcommittee of the U.S. House of Representatives,[11] Senator Wendell Ford's subcommittee of the U.S. Senate,[12] and in testimony before the Energy Commission of the state of California and the Energy Research and Development Administration of the federal government. Those economic projections have always been on the cautious side, assuming high lift costs for the space manufacturing equipment, large mass for the SSPS plants, relatively low productivity in space, high interest rates on investment, and low electric rates for SSPS power supplied to the Earth. Yet all the projections confirm that SSPS plants built at a space manufacturing facility out of nonterrestrial materials should be able to undersell electricity produced by any alternative source here on Earth.

The most recent studies, based on the "Low Profile" approach to space manufacturing, appear even more attractive, because they indicate useful production starting well before Island One is built, at a time when the total investment is much less than the $100 billion estimate originally made for Island One.

By now our planning group benefits from the advice of senior executives in the electric utilities and investment communities. From them we have learned a good many realities that help us in guiding our research. For one thing, it seems almost certain that we cannot expect private capital to invest in space manufacturing until the risks have been reduced almost to zero. Government funding, possibly by a consortium of several governments, will have to carry the program at least until a pilot SSPS, not necessarily made from lunar materials, has supplied energy to the Earth. At the same time, we will have had to demonstrate that we can bring out lunar materials and process them in space, to get the same elements used in building the SSPS. Above all, the economic studies made at that time will have to show that SSPS power can undersell all competition. Once that happens, though, it appears that private capital in large quantities should be available for the expansion of the program to full capacity.

As of the late 1970s, the lowest-price electric power in

the U.S. costs around two cents per kilowatt-hour at the power-plant. Our goal is to undersell that price, whether or not the prevailing rate goes up in later years.

As the possibility of construction of Island One is examined with greater care and in more detail, both the engineering and the economics can be studied in far greater detail than they have been. The most significant point about this discussion, though, is that already it can be carried out at the level of engineering and economics. There is no dependence on any basically new physics, nor on any great extrapolation of engineering practice beyond what is customary today.

One of the graphs prepared for the evaluation of space-manufactured SSPS plants was given in congressional testimony (Appendix II). According to that graph, rather quickly, within thirteen years from the initiation of heavy investment in Island One, the rate of construction of new generator capacity in space could exceed the annual growth needs of the United States. Not long afterward, the total energy so far supplied from space could exceed the total stored in the Alaska North Slope in the form of oil.[13] The contrast is glaring: in the case of Alaska, at that time there would be little remaining from the Alaska pipeline (except perhaps for oil slicks on the water and some discontented elk) while satellite space power could still supply clean electrical energy to the Earth for another five billion years—the estimated life of the Sun.

For an enterprise demanding investment capital at the start, and yielding profits at a later time, economists calculate what is called the "benefit/cost ratio." Taking account both of interest charges and of inflation, the benefit/cost ratio summarizes whether a possible invest-ment is worthwhile or not. Even without the cost-savings of the "Low Profile" approach, the benefit/cost ratio for space manufacturing is much above one, indicating that in spite of high interest rates and low power charges, the Island One program would be a paying proposition. To get that favorable result, it appears that exponential growth of the manufacturing capability in space is very important; a slower, linear growth doesn't pay off fast enough to make up the interest charges on investment.

When the amortization of power plants is complete, the cost of power generated in this way should go down, because the satellite stations should require little maintenance and will be using free energy—given by an efficient, clean thermonuclear reactor which has been located for us at a comfortable distance of 150 million kilometers.

If this development comes to pass, we will find ourselves here on Earth with a clean energy source, and we will further improve our environment by saving, each year, over a billion tons of fossil fuels, now lost to heat and smoke in driving our electric generators. Given a worldwide market which may be several hundred billion dollars by the year 2000, probably the industries at L5 will grow rapidly in numbers and size, to satisfy so urgent a demand.

If satellite power stations are built at L5 rather than on Earth, there will be important environmental consequences. For every SSPS that would have to be lifted from the Earth if built here, many times as much weight would have to be dumped into the atmosphere in the form of rocket fuel exhausts, to lift the SSPS components. The total quantities run in the range of hundreds of millions of tons per year, if SSPS power becomes dominant in the world economy. No one knows what the environmental effects of those exhausts would be, but it seems sure that writing the "environmental impact statement" for such a program would be no easy task. By contrast, the establishment of space manufacturing requires only about a hundredth as much lift tonnage from the Earth, and is within the existing space-shuttle traffic model, whose impact has been carefully studied and shown to be safe.

A major open question, of course, is what fraction of the mass of an SSPS power-plant couldn't be obtained from nonterrestrial materials, and so would have to be lifted from the Earth anyway. If we were using asteroidal materials, we could be sure of having in quantity all the elements we have on the Earth. The Moon, though, is poor in hydrogen, nitrogen, carbon, and some heavy metals. Fortunately NASA has now begun to study that question, and over the next years we may hope to see designs for satellite power stations optimized for the use of lunar rather than terrestrial materials.

So we're beginning to perceive a possible branch in the development of satellite power. The best game-plan seems to be to keep the options open: build small pilot-plants, improve solar cells, and meanwhile push the early research and development of mass-drivers and lunar-soil processors. After a few years of research, when the numbers are clearer, there will then be a point where a rational decision can be made, either to develop the huge, very advanced lift rockets needed for a satellite power system built on the Earth, or to put a similar amount of money into developing the "nonterrestrial alternative."

If the efficiency of industry in space improves to the extent predicted by some students of the subject, the cost of satellite solar electric power delivered on Earth could drop to much less than one cent per kilowatt-hour, at the antenna on Earth. If that occurs (I am not yet willing to claim that it will, because research has not yet been done in sufficient detail) it would have profound consequences for international politics. With low enough electricity rates, it would be possible to synthesize clean artificial fuels, which could compete economically with gasoline and render this and other participating nations independent of oil imports.

It would be well within the capabilities of Island One to produce a large optical telescope, made up of many individual mirrors. Great resolution could be achieved by locating the individual elements of that telescope in a precise array stretching over a considerable distance, instead of combining all the mirrors together.[14] In designing such a system it is natural to consider linking the mirror elements by a mechanical structure, but that might be the worst possible thing to do. A mechanical link would expand and contract with temperature changes, altering the mirror spacing. It might be preferable to take advantage of a zero-gravity location by building a large number, perhaps several thousand, of individual glass mirrors, each a meter in diameter, and providing each with a small

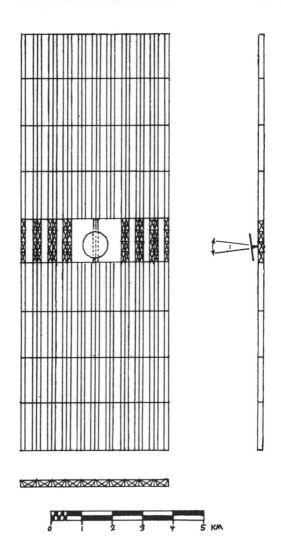

0 1 2 3 4 5 KM

Satellite power station,
nearing completion at
Island One. Central
antenna sends power to
receiver on Earth.

locator-module, equipped with station-keeping gas jets. The heavy parts of such an array might be made at the space community, while the light, complicated, labor-intensive parts would be brought up from Earth.

If the elements were linked only by light beams, their spacing could be established by the unvarying number of wavelengths of light between each pair. That nonphysical linkage, computer-controlled, would have the further advantage that the mirrors could be programmed to separate and reform, like dancers in a slow-motion ballet, according to the needs of a particular astronomical experiment.

If located in a cross-shaped array, with individual elements spaced ten meters apart, a telescope of that kind would have the theoretical capability of resolving something as small as a changing weather system, one thousand kilometers on a side—on the planet of a star ten light-years away!

Once Island One is in full operation, almost surely the scientists will argue strongly to have part of its production capacity put into ship-construction. Even a percent or so of L5's output of aluminum, magnesium, titanium, iron, and other useful metals will be enough, over a period of two or three years, to build a large research vessel which could be in many respects a space-borne equivalent of Darwin's ship the *Beagle*. Equipped with an engine which might be a slimmer, longer version of the lunar mass-driver, this research-spacecraft could voyage to an asteroid, using as reaction mass crushed rock dust. The *Beagle II* might have a crew much larger than the original H. M. S. *Beagle*'s fifty, and they would form a small, self-sufficient laboratory village. The "launching" of their craft would require none of the flame and thunder that accompanies a launch from Earth. Instead, floating at rest in space at the entry dock of Island One, the vessel would seal its entry port and quietly cast off. Let us imagine the voyage as though its details were certain:

When power is fed to the engine, the ship will begin to move almost imperceptibly, hardly an arm's length in the first half minute. But a day later it will be only a small dot of light in a telescope, and after a month it will be ten times the distance of the Moon.

When its crew, many months later, performs a rendez-vous with a small asteroid, the scientists aboard will take as much time as they like to study the planetoid in great detail, measuring its mineral content, assaying its resources of carbon, nitrogen, and hydrogen, and collecting tons of samples. Much of their work will have direct "applied geology" applications, to the later use of asteroidal material for construction; that is probably what will pay for the trip. Other work, carried on a fraction of the time, may seem then to be without direct application; the view from later on, a few years afterward, may be quite different.

While the scientists are at work, the engineers will use their on-board machinery to mine and collect several thousand tons of rock and dust as reaction mass: fuel for the next leg of their voyage. When the voyagers cast off and feed solar power to their engine to return or go farther afield, much of the scientific information collected will already have been analyzed and radioed back to L5 and to Earth. During the long months of the voyage that follows, samples will be analyzed in the ship's laboratories, information digested, and scientific papers written. Submitted from deep space by radio transmission, these papers may bear such identifiers as: "Carbon 12/Carbon 13 Analysis for Asteroid 2655; by ———, *Beagle II* Research Laboratory, en route to Ceres."

As the voyage proceeds, a small on-board rock-crushing plant will run continuously, providing rock dust as reaction mass for the engine. Unless the crew finds it too confining to be isolated in a small traveling village, a vessel of this sort could cruise among the asteroids for years. Surely families will travel together, and children will go to school on the *Beagle II*, sharing work and relaxation with their parents. Later, in the days of Islands Two or Three, much larger ships can be built, carrying with them to the outer regions of the solar system whole research institutes and sections of universities.

It seems likely that Island One will become a favored place for scientific sabbaticals from Earth. Especially for young scientists, not yet concerned with marriage and family, the opportunities for research at the space habitat in radio and optical astronomy will be unexcelled. A cycle may well be set up in which a scientist will arrive for a year of intensive data-taking, then exchange his place with a new arrival, while he returns to Earth to analyze his data and write his conclusions in article form for the scientific literature.

For research in radio astronomy, most antennas are arranged in geometrical patterns, like crosses or circles. One special type of antenna, though, might be formed as a huge parabolic dish. I confess to some misgivings about the use to which this great mirror would most likely be put, yet I cannot deny that Island One would be the ideal place for its construction and use. This antenna would be used for a project known as "Cyclops"—the big eye; a search for extraterrestrial civilizations.

For more than fifteen years there has been interest in the possibility that there may be other intelligent civilizations in our galaxy, which could be members of what some people have called the "Galactic Network."[15] It is difficult to say, on the basis of any theory, what the chances are that such civilizations do now exist. The idea that we, as (to some degree) intelligent life are unique is of course absurd: the more we learn about the origins of life, the more we realize that the conditions under which life first began on Earth must have been duplicated many times over in other parts of the galaxy. The crucial unknown quantity, though, is outside the natural sciences entirely: it is the lifetime of a communicating civilization.

Our galaxy is disc-shaped and has a volume of a thousand billion light-years. Many of the individual stars within it may live in a stable manner for several billion years. In the modern view, perhaps one in ten of the hundred billion stars of our galaxy may have planets, and so be "likely" places near which life may originate. In 1959 Phillip Morrison and Giuseppe Cocconi speculated on the possibility of searching for extraterrestrial life with the sensitive receivers used in radio astronomy.[16] Soon afterward, Frank Drake carried out the first search intended specifically to look for such intelligently directed signals; his "Project Ozma" was capable only of examining a few nearby stars, and found only natural signals. [17]

Those scientists most interested in the search for intelligent extraterrestrial life recognized some time ago the importance of two vital numbers: the odds that such life will develop on a planet of a "likely" star and the length of time that a civilization will be engaged actively in radio communication. The importance of these two numbers can be illustrated by examples: If life in the galaxy is very abundant, perhaps as many as one in ten of all stars with planets become the nurseries of new civilizations at some time in their evolutionary history. If so, there may be as many as 100,000 stars within 1,000 light-years of our Sun, each of which becomes at some time the birthplace of a civilization. What is the chance that we, searching all the one million planet-bearing stars within that great sphere, will find at least one which is beaming signals toward us? That depends very much on the second critical number: the duration of communication. Even if the average civilization remains actively engaged in communication for 100,000 years, and even if it devotes to that purpose an effort sufficient to beam signals continuously toward every likely star within its own 1,000-light-year "sphere of interest," the chances are only about even that we are on the scene at the right time to receive an intelligent signal. The reason is that for any given civilization the period of communication corresponds, in our example, to a brief moment of time which occupies only one part in 100,000 of the whole evolutionary history of its star. [18]

The uncertainties connected with such numbers are so great as to leave open two extreme possibilities: first, that communicating life is sparse, that the duration of communication is rather short on the galactic time scale (by short, I mean 100,000 years or less) and that we are, therefore, *at this moment* alone within a 1,000-light-year distance, or even alone in the galaxy.

The other extreme case, still open as a possibility, is that the galaxy teems with communicating life, that the duration of the "attention-span" of civilizations is many bil-

189

lions of years, and that consequently as soon as we put our ears to the ground we will hear the beat of distant drums.

With so much room for the imagination I find it irresistible to add my speculations to those of so many others who have written on this topic. My guess, and it can be no more than that, goes this way:

First, I think that soon after a civilization reaches our own modest level of technological competence it becomes unkillable in the physical sense; the reason is just the topic of this book: the movement of life into space. As R. N. Bracewell has written:

"When we have colonized interplanetary space—which could be early in the 21st century, according to Princeton physicist Gerard K. O'Neill's timetable—we will have concomitantly achieved independence of the terrestrial catastrophes that lie ahead. Survival of the fittest, on a time scale of geological upheaval, may mean that communities over a certain age will be those that have succeeded in colonizing space." [19]

I would add a remark to Professor Bracewell's comment. Freeman Dyson has pointed out that there may well be very intelligent civilizations which have no interest in technology. I quite agree, but would guess that any civilization which becomes interested enough in the natural sciences to develop radio astronomy will achieve, almost at the same moment in its evolutionary history, liberation from its parent planet. Logically, then, I do not believe that war or natural catastrophe will constitute, in many cases, the limits to the duration of a civilization capable of communication.

I do have serious reservations about the probability that a civilization capable of communication, and stable enough to have a long lifetime on the galactic scale, will in fact choose to communicate. I readily concede that my reasons could be excessively anthropomorphic. They are closely connected with my misgivings about the entire concept of Project Cyclops.

On our planet we have seen, again and again, the effect of the contact between a primitive culture and a more advanced one. Almost invariably the more primitive is shattered. The destruction may not be intentional; often it may not even be physical. Yet it occurs, because the values and the knowledge gained over the centuries by the primitive civilization become, overnight, of little value in comparison with what is available from the more ad-vanced.

When I have considered the effect of our discovering, one day, signals from a more advanced civilization (note that it would be, almost certainly, millennia more advanced than we are because of our own position at the threshold of communication) it has seemed to me overwhelmingly probable that the first effect of the discovery, as soon as the excitement and the novelty have worn off a little, would be to kill our science and our art. What purpose to study the natural sciences? We already know that they are universal, so if a civilization now radioing to us is as many thousands of years ahead of us in knowledge as we are from the Neanderthal, why continue to study and search for scientific truth on our own? Gone then the possibility of new discovery, or surprise, and above all of pride and accomplishment; it seems to me horribly likely that as scientists we would become simply television addicts, contributing nothing of our own pain and work and effort to new discovery.

In the arts, music and literature, the case may be some-what more unclear; yet on Earth the almost invariable consequence of contact between a primitive civilization and one more advanced is the stagnation of the arts in the former. Only in the form of a "tourist trade" does art survive, in most cases.

If this sequence of effects is of more than local signifi-cance, as I think it is, it will be quite obvious to any civilization more advanced than our own. I would then add one more assumption: that the same characteristics which render a civilization immune to intellectual decay and stagnation, if there be such characteristics, are accom-panied by a repugnance to inflict harm on others, in particular to other "emerging" civilizations more primi-tive than its own. In that case, "They may be out there, but they're kind enough to keep quiet."

If civilizations combining great age, great social stabil-ity, and a continued lively intellectual interest do exist, and if those characteristics are accompanied by a concern for the development of primitives such as ourselves, is there any kind of signal that could be sent out to us that would carry great potential for good, and little for harm? Perhaps there is: the flash of a lighthouse, a simple message endlessly repeated, carrying just enough information so that we know it was formed by intelligence. The simple fact of its existence, proclaiming "you are not alone," could be of great help to us in our darkest moments. It would pull us outward, and spur our development; after all, we do not wish to appear as country bumpkins when true contact is finally made. At the same time, after the ten-thousandth repetition of the same brief message, it will be clear to us that we must continue to gain our knowledge of the universe step by painful step, through our own efforts, and that physical travel to great distances will be needed before we answer the question: are they still there, or do we hear only the echo of a civilization that vanished long ago?

Proceeding now to the most speculative assumption of all, I consider that this age of natural science, in which we now find our own human civilization, may be a relatively brief epoch in the history of a long-lived species. We are in the midst of a knowledge explosion, and if our rate of acquisition of new scientific knowledge continues to accelerate, as it is now doing, it seems to me quite likely that within much less than a thousand years we will know, if not everything about the natural world, at least so much that science will no longer be of great interest and challenge. In that case I would expect that our most talented individuals, a few of whom now study the natural and biological sciences, would turn their attention to the arts, or to the greatest intellectual problem that is now imaginable to me: the riddle of consciousness. My picture of an advanced civilization is one in which science, aided by computers with an intelligence level far higher than that of any living being, will already have answered all the merely physical questions. Some individuals may take part in direct exploration and exploitation of new star systems, slowly spreading the culture of their species in an expanding sphere from their parent star. I consider it probable, though, that in the advanced stages of a long-lived civilization the physical world will be taken for granted, as something long since understood and thoroughly tamed. Most of the interest and activity, I would guess, will be intellectual, artistic, and social.

After so much that is speculative, it is almost a wrench to return ourselves to the "little" world of our own solar system and the few decades immediately before us. We do so to consider the practical question which remains when debate about the value of Cyclops has gone on long enough: if it is going to be done at all, what is the best and most economical way to do it?

The answer to that question seems rather clear. Cyclops, in its original form, was studied by a group of two dozen people during the summer of 1971, at the NASA-Ames Laboratory in cooperation with Stanford University. The leader of this group was Dr. Bernard Oliver, of the Hewlett-Packard Corporation, and the result of the study is a thorough, excellently prepared report entitled "Project Cyclops."[20] The report concluded with a proposal to construct, somewhere in a lightly inhabited desert region, an array of up to a thousand radio telescope antennas, each of large size, all of which would be steerable so as to point in a fixed direction as Earth turned under them, all braced against wind and storm, and all tied together electronically to function as a single giant receiver.[21] The total cost of the effort, if in fact all the antennas were built before an intelligent signal were received by them, was originally estimated as fifteen billion dollars; it could be reduced if advances in receiver sensitivity allow the same result to be achieved by a smaller array of antennas.

As an exercise, I looked into the question of building the equivalent of a Cyclops array as one of the early tasks of Island One. The space-borne Cyclops would be far simpler; probably a single giant parabolic dish antenna, five kilometers across, located at a short distance from the space-community. It would require only a single receiver system, which could easily be updated to remain at the summit of the electronic art as the years of search went on. The problem of noise arising from the many communica-

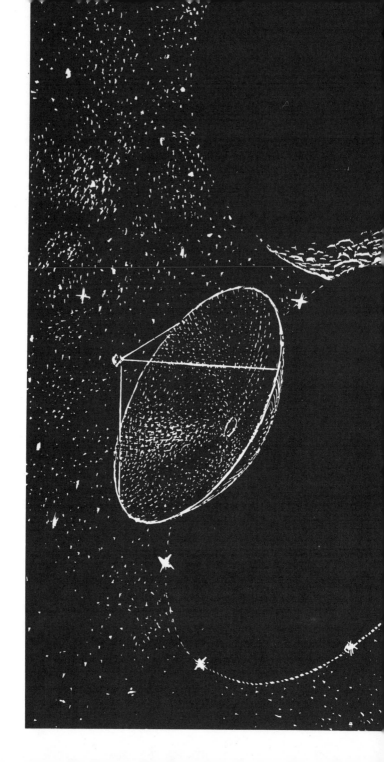

Cyclops dish antenna,
several kilometers across,
and its metal shield to
block interference
from Earth radios.

tions transmitters on Earth and in space would be overcome by the simple expedient of locating a disc-shaped baffle, twice the size of the antenna, a short distance away.

For operation in the zero-gravity, wind-free environment of space the antenna and its noise shield could be light in structure, composed (in my estimate) of a geodesic frame covered by a thin skin of aluminum. The total mass including the shield would be hardly a tenth as much as that of an SSPS. Assuming that all of the complicated machinery (electronics, motors, motor drives, and so on) were brought up at high cost from Earth, and putting in generous figures for the costs of fabrication and assembly, the total expense for the Island One Cyclops still comes out only about one tenth to one twentieth of the cost of an equivalent installation on Earth. The L5-Cyclops would have a further advantage illustrated by an amusing little speculation: suppose that among the one million stars that are searched during a thirty-year period there is in fact one which is beaming signals toward us; suppose further that the "program" has a duration of many years. After all, the beings at the sending end might be far more long-lived than ourselves, and they might have a lot to say. Once we lock in on the signal, the L5-Cyclops can continue to point at the right place for as long as the program lasts. In contrast, a Cyclops antenna array on Earth, on the Moon, or in low orbit would be blocked half the time from receiving the signals. If that were the case, we can imagine the resulting Congressional investigation:

Senator X: "Do I understand, Professor, that we are missing half the program the Arcturians are sending, and that you are proposing that we now build a new antenna system at L5 in order to replace the one in Nevada?"

Professor Z: "Yes sir, that is quite correct. Of course, when the Cyclops program was initiated, we did not anticipate receiving signals with this particular time structure."

Senator X: "Are you trying to tell me that when you came in here to ask for fifteen billion dollars, you weren't anticipating the possibility that your search might succeed?"

196

We will leave Professor Z in his rather sticky situation, and turn now to another application of the industrial facility at L5.

If the calculations I have described are not wildly off, the work force at Island One will be in a location so favored for industry that there will be strong pressure to enlarge the "beachhead in space" by constructing larger habitats.

Whatever group builds the first community, the success of Island One will prompt others to share in the earnings of L5 industry. Even on a three-shift basis and with a population of which most people will probably be among the work force, the first Island One will not be able to satisfy, by itself, the demands which an energy-hungry world will make of it. Even while the first few communities are being built, their designers (or possibly an entirely different group) will be planning for the next step in size: Island Two. The choice of size for the new generation of space communities should be made carefully, because for lowest cost it will be best to choose an optimum size and then replicate it in large numbers, using automated machinery and dry docks all suited to one set of dimensions.

Island Two should be large enough to form an efficient industrial base, but small enough so that transportation within its valleys will be easy, and so that its government can be simple and nonbureaucratic, functioning with a minimum of red tape. At a rough guess, the space-residents may be ready to tackle something the size of Island Two after there are already a dozen or so communities of the size of Island One.

For economy of the structure which must contain the atmosphere, the internal pressure may be chosen similar to that in a high-altitude town on Earth: Denver or Mexico City. Much calculation will be required before we know the optimum size for Island Two, but my present guess would place it not far from an 1,800-meter diameter (about 6,000 feet) with an equatorial circumference of nearly four miles. Island Two could house and maintain a population of 140,000 people, possibly in a number of small villages separated by park or forest areas. Each such village could be similar in size and population density to a small Italian

197

hill town. As a former resident of such a community, I can confirm with nostalgia that it is one of the pleasantest arrangements for living so far developed on Earth.

Any comment I make about the city architecture and geography of a space community is, of course, no more than a guess. It may well be that several different types of arrangements will be worked out, perhaps even within a single habitat, so that without leaving the habitat people can enjoy the variety of "evenings out" in villages quite different from their own.

As in the surroundings of the first habitat, all the heavy industry of Island Two will be located outside, at least a few hundred meters away, in zero-gravity.

Even while the first islands in space are being built, work will go on to upgrade the lunar mass-driver, for the task of increasing production of export products and of additional communities. A solar power station might be located on a mountain peak at the lunar north or south pole, where sunshine would be available full-time. A transmission line from the pole to the lunar mine would allow the mass-driver to double its throughput, without any change in the machine itself.

By the time construction of Island Two begins, there may be more than one mine on the lunar surface. Perhaps by then there will be a small industry on the Moon for building mass-drivers and their solar power supplies. In the long run that will be the way to reduce shipping costs to a very low value (a few cents a kilogram). By then, too, we may be exploiting the vast reserves of the asteroids, and not long afterward, if the economics is favorable, we may shut down the lunar mines, and leave the facilities there as ghost towns.

The quite conservative economic scenarios that were developed early in our study of space manufacturing were based on a doubling-time of about four years for the number of Island One communities, so that after about fifteen years the population in space would be over a hundred thousand. That number should be in the right general range to satisfy all the U.S. demands for new generator capacity soon after the turn of the century. It seems likely, though, that by then if not earlier the space communities will be responsible for supplying new generator capacity to all nations in need of it. As a rule of thumb, an Island One community, fully employed in heavy industry, could produce about 200,000 tons of finished products each year; more than two power stations, if it had no other employment. World needs by the beginning of the next century may be as high as fifty or more big SSPS stations coming on line every year, so the time may be not many decades off when the population in space may exceed a million people.

If automation is carried far, so that repetitive operations are done by a small work force, the replication time for habitats even of Island Two size could be as short as two years. The conditions at L5 seem made to order for such a development: zero-gravity for assembly of large objects by lightweight machines; no weather, so computerized production will not have have to cope with the vagaries of seasonal variation and natural hazards; unlimited energy, and a task which consists of the repetition, thousands of times, of the same assembly operations with identical simple structures.

If the fastest possible time scale is reached, fifteen years from the beginning of construction there can be many communities at L5, with several hundred thousand people living and working in space. I hope that they will include young and old, children and the elderly, as well as people of working age. During those years, the sale, lease, or donation of Island Two structures as "turnkey" industrial facilities appears to me a likely possibility. The cost of such a habitat should be not much more than that of the original Island One, because by then the work force at L5 will be quite large enough to produce every necessary variety of machine and part for construction; only liquid hydrogen and possibly nitrogen or carbon will still have to be brought up from Earth.

For an underdeveloped nation or consortium of nations, the period during which several Island Two's are built will almost certainly be a time of excitement and opportunity. For a nation of a billion people (within two or three decades there will be at least two nations of that size) a space community for 140,000 people could then be bought, over a ten-year period, for a cost equivalent to a few dollars per person per year. As a beachhead in space, from which further rapid expansion could then take place without additional foreign capital, such a community could be an attractive investment, expecially when the possibility of its exponential growth is considered. It is speculation, and possibly nonsense, but thought-provoking that with a replication time of two years for new habitats, a nation of a billion people, which purchased an Island Two Structure, would obtain in just eighteen years a growth rate of new lands in space fast enough to absorb a population growing even at 4 percent per year. Later I will explore that possibility further; for the moment, though, let us turn to what life might be like for the pioneers who may settle Lagrangia.

10
TRYING IT OUT

During the settlement of our own New World here in the western hemisphere, communications across the sea between family members were very important. Letters from the early immigrants allayed the fears of relatives left behind, and in many cases encouraged them to follow. In the settlement of L5 the communications with the "Old

Countries" will be far more rapid: television phones can operate with a time lag of less than two seconds. It seems likely that even the earliest space communities will be equipped with electronic mail-transmission systems, and I suspect that letters between family members will be as important for the humanization of space as they were to the settlement of our country. When time does not press, and there is a need for the comfort of handling the actual piece of paper touched by the sender, mail sent on a space-available basis may be more satisfying.

Here are letters of a kind that might be written by people emigrating to L5 a few years after the first pioneers. Unlike those youthful emigrants who might be in the majority, this is imagined to be a couple whose children have grown up, married, and established families on Earth. Work experience and a record of stability and responsibility could be important factors influencing the "selection committees" which will play an important role inevitably in determining who goes to the early communities, though with the passage of time it can be expected that eventually most of the people who may wish to go to L5 will have the opportunity of doing so.

Dear Peggy and Arthur:

Jan. 15, 20–: Jennie and I have been in Station One for twenty-four hours now—I'll send off this note by the video-mail while our impressions are still fresh. We were glad to get away from the slush and the wind up North, but even Cape Canaveral was pretty cold, and we heard that they were getting worried in Florida about the orange crop. Once in the Space Terminal it seemed like familiar ground to us, because of our six months at the Training School. Some of the people from our class were going to be on the same shuttle flight, too. After all the ticketing, the last medical checks, and getting our personal kits weighed, we went through to the locker rooms where we had to say good-bye to our clothes. Then showers and hair-washing, and out the other side to the "clean rooms"; nobody wants to give a free pass for L5 to any plant-eating bugs. Our space clothes were waiting for us, all clean and pressed. Of course we'd tried them on at the school, and

202

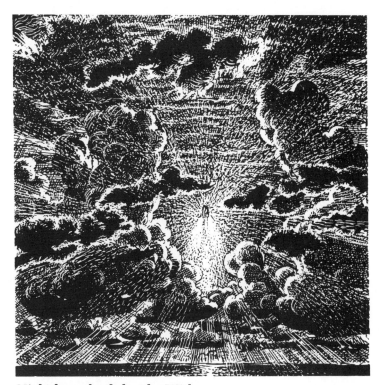

Night launch of shuttle. With
solid boosters and main engines
burning, Orbiter lifts through
first cloud layer.

Jennie'd sent hers back a couple of times to get the fit just
right. I don't blame her—these light clothes don't leave a
lot to the imagination.

The shuttle was already on the pad when we came into
the waiting room; the spaceport crew was fueling up. We
waited about an hour, but didn't call you—nothing new to
say yet. Then the 150 of us filed on and settled into the
bunks—the cushions were thin, but we knew we'd only be
on them for half an hour. On the TV panels we could see
our own lift-off, and it sure felt different knowing we were
on top of those fireworks! The g-forces weren't bad, especially lying down; just like the centrifuge at the school. At

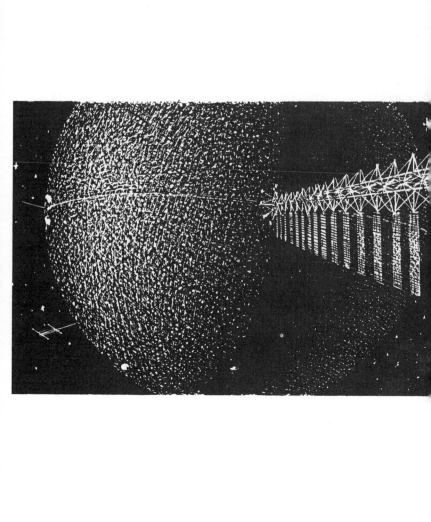

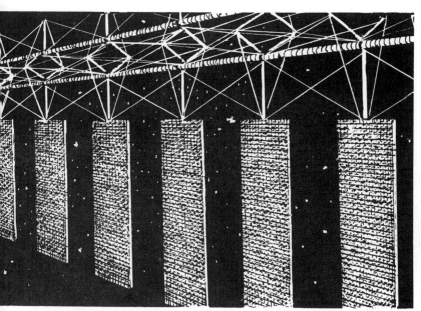

The *Robert H. Goddard*,
a space-vessel for two
thousand passengers.
Mass-driver engine
is powered by
solar-cell arrays.

the end it was about three gravities, and I could still lift my leg without much trouble. Zero-g was very strange at first, but we kept still like the book says, and didn't get sick. On the TV we could see the shuttle moving up to Station One, and we felt the bump when we docked. The station hostesses floated in and helped us out and into the station—that took a while, about another twenty minutes I guess. Altogether, from lift-off to Station One was less than an hour.

They have a ramp leading "down" to the outer rim, so as we walked down gravity built back up to normal. The Station One lobby and restaurants have been on TV so often I won't say much about them, but I do want to tell you about the people. We were pretty lucky—there were only another 24 hours to go in the three-day cycle between ships, so the station was pretty full. The seven shuttle flights before ours had brought up groups from a lot of different places: Chinese, Russians, a fair number of Indians, and one Nigerian group. From where I'm writing I can see Jennie—she's in one of the garden rooms and seems to have struck up a conversation with a girl who looks to me as if she came from someplace in South Asia—I guess they're both flower-crazy.

Jan. 17: By the time all 2,000 of us were in the station the hotel was pretty crowded. It's good though that they have so many observation windows—we spent most of our time just goggling at Earth. I took a lot of color slides, because it may be two or three years before we get this view again. The hotel rooms were nice enough, but we didn't spend much time in them: too much to see, with Earth and the continuous movies they run in several theaters, and all the strange people.

We went to our room to see the *Konstantin Tsiolkowsky* come in—the view was a lot better from the TV. It was a pretty sight: first the end of the engine came in view, with its bright searchlight lighting up the clouds of vapor as they came shooting out. We could almost count them—like a movie running a bit slow. It took a long time for the whole length of the engine to go by; the long straight mast, and the yardarms with their flashing red running lights. We couldn't see the guy wires that keep the whole thing straight. Then the ship herself came in sight: just a big ball, with no windows at all—like the head of a tadpole, and beyond it a big dish reflector for solar power. It took about three hours for all of us to file on with our kits, and get settled in our cabins. We're not used to zero-g yet.

The captain gave us a nice speech over the video while the *Tsiolkowsky* got under way. Told us about how the passengers and crew are all on three shifts, matching three time zones on Earth eight hours apart: Moscow, Cape Canaveral, and Western Pacific. So the restaurants are going pretty much full time. There aren't any windows, of course, because of the cosmic-ray shield, but the big video panels in every room give us good views, and they've got it set up so the cameras are fixed—you couldn't tell by looking at the video that the *Tsiolkowsky* is rotating.

Jan. 18: They sure keep you occupied here. I can see why the captain calls the *Tsiolkowsky* and the *Goddard* the "Flying Schoolhouses." Jennie and I are in classes for brushing up on our 800-word Basic Russian and Basic Japanese, and they've got a kind of nice arrangement about the meals. While one shift is having breakfast the other is having supper, and at that meal they use place cards. It's pretty clear what they're trying to do: sitting at a table for four, with a couple from maybe Russia or Japan or China, you can hardly help getting to know people. The Japanese couple we met this morning are in power-plant construction, like us: he's an expert on casting titanium turbine blades. Since Jennie's been training for the last half-year to be a blade inspector, they had some shoptalk to get into. The Japanese girl is an agricultural specialist, so I learned quite a bit about how they get so darned much food out of a little land up in the Japanese communities. I'll have to admit, though, that their English is a whole lot better than my "Basic Japanese." I think they're cheating—they use a lot more than 800 words!

There was big excitement early today when we passed the *Robert H. Goddard* on her way in toward Station One. She was in view for more than an hour, and our crew gave us some nice telescopic views, with enough warning so that there was a lot of camera-clicking. The camera population must be more than the passenger list.

We met an Indian couple at dinner tonight. He's in construction, which makes sense I guess because the Indian government is concentrating on a fast buildup in the number of their habitats, rather than building mainly power plants like everybody else. We passed the orbit of the power stations on the first day—guess I forgot to say. Every now and then, as we spiral out, we get a bright glint off one of them, far in close toward Earth.

The L5 communities are getting visibly close now, and everybody is pretty excited. I've got to admit I get some butterflies sometimes: this is a young crowd, mostly, and even after all those tests I wonder sometimes whether Jennie and I, at fifty, are still good at learning new ways. So far we've liked it, I'll say that. The captain, in his little daily speech, is pretty funny; I guess he's developed quite a line of patter, after making the trip once every twelve days for a couple of years. I don't see, though, why he keeps apologizing about the food; it's a good deal better than what the airlines give you, at least. Today Jennie ordered a curry from the Indian menu. I copped out—took a ham steak, but I tried a bit of the curry and liked it.

Jan. 20: It was really great having that long video-call with you this morning. That free half-hour every week is going to mean a lot to us. Seemed as though the grandchildren had grown even since we left. Of course, we forgot most of what we wanted to say, and so much has happened that we couldn't have squeezed it in anyway.

We told you they docked us at Island One; they seem to use it partly as a receiving hotel now, a place where the *Tsiolkowsky* and the *Goddard* can dock, and where people get sorted out and sent to the communities they'll be living in. We exchanged some names and addresses, and already have some invitations to go visiting when we all get settled in a bit.

Island One is small, of course—only 500 yards in diameter. This one runs at Canaveral time, and two other communities of the same size, nearby, run on the other time zones. Many of the people we met on the ship landed at the others, so as not to be time-shifted on arrival.

I wonder what it was like, for the people who lived in this first habitat for several years, before the first of the

208

Island Two's were done. Not too bad, perhaps. Jennie and I were given one of the smaller apartments, two big rooms with kitchen and bath, with a nice garden. This first Island One runs at a constant Hawaiian climate, because they weren't sure about temperature changes in the beginning, and didn't want to take chances with the structure. The old-timers say the climate in One is dull, but after Michigan in January we're happy to soak up the sunshine for a while. The garden has some big tropical flowers, and I can see that the people who lived here first liked avocados; they had several trees, and one was just right to give us a fresh avocado with our lunch.

It really is sort of a vacation atmosphere—the good weather, and so many new things to see. Of course, we had to try Island One's first new possibilities: human-powered flying and the slow-motion diving.

On our deck chairs in the garden, soon after we settled in (I won't say "unpacked," because with a 100-pound baggage limit, we didn't have that much to unpack!) we could look up and across at the big corridor that leads to the "ag-area"—the part where they grow the crops, using all mechanical equipment. The curved surface of the habitat is all terraced and planted—a mass of green and bright colors. The Sun is at an angle of about 11:00 A.M., all the time except when they shut it out with some screens outside, for nighttime. Each morning around seven there's some "rain," so when we wake up everything is fresh and there's a nice clean smell of rain and flowers in the air. Island One is too small to have any real weather of its own, though, so the "rain" just comes from some spray pipes that we can see, seven hundred feet above us, when we look hard.

Exactly above us we can see the gardens of the apartments on the far side, and then the curve of the sphere. For some reason, it doesn't seem so strange to us to have trees growing straight down as it does to have them seeming to come out horizontal, as they do from the gardens a quarter-circle away from us.

Many of the gardens are open, but someone told us that the settlers who prefer the small size of Island One, and have stayed here rather than move, have little gauze

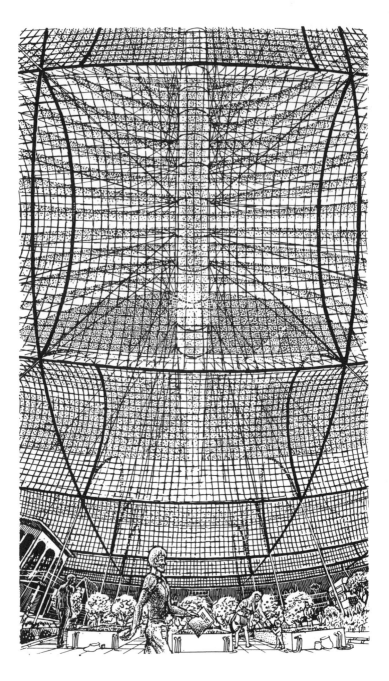

sunshades over part of their grass, so that they're able to sunbathe in the nude and still not be seen from the "sky." As we look up we can see the figures of people flying, 800 feet above us.

The apartments here are grouped in terraced buildings, so that each can have its garden, and the buildings are set into little villages, separated by trees and parks. In our exploring we're getting our exercise, because there are pathways, not roads, and most of the villages are at least partway up the slope from the "equator."

People seem to love flowers here; all the paths are lined with them. I guess it's just so easy to grow them—no weeding, no bugspraying, and just the right amounts of rain and sun. I understand that in each community the Garden Club is one of the most important organizations there is, and that particular people volunteer to take care of individual little areas of the paths and parks.

Down near the river there's "Fifth Avenue" where nearly all the shops are. It's on two levels, with wide sidewalks and a lot of planting. About half of it must be little restaurants—it seems that when Island One was new everyone was working so hard, men and women both, that a lot of them didn't do much cooking. The restaurants are all small, and many of them are buffet-style, most with salad bars. There are numerous bookshops, and a main library, and quite a few small cinemas.

Still lower down, past a belt of trees, there are tennis courts and the playing fields, and of course the park and beaches at the equator itself, by the river.

On our first exploration we kept looking up and seeing people flying, and it looked too good to miss, so we began walking up through one of the villages, and beyond it to where the hill gets steeper and steeper. The sensation was really strange, because as we climbed we got lighter. Past the green park area we were on one of the bridges across the windows, climbing at more than a 45-degree angle, but finding it easier because of weighing less. Beyond the windows we were on a very steep, winding path, through

Orchards and farm buildings in
agricultural areas of Island One.

211

almost a jungle of ivy and shrubs, like a Hawaiian hillside. Up at the very top we found a number of people, because all of us newcomers were exploring at the same time. I'll have to admit that as soon as I tried flailing around in the zero-g clubroom (it doesn't rotate) I started feeling a bit ill. Jennie loved it, though, and when someone had a pedal-plane free she was the first of us to try it. I watched her from the zero-g room; the pedal-plane sets you at an angle almost like lying down, and there's only a bar at waist level—no real seat. The wings are small, but there are three sets of them; it's a triplane. The two big propellers are almost as big as the wings, and go in opposite directions when you pedal.

Jennie had some problems right at the axis, because there wasn't any "down" and the plane was designed to work in at least a little bit of gravity. Once she moved down a few feet, however, she was ok and pedaled out. She stopped pedaling and drifted into the thin netting that's out there, pedaled for another quarter-mile, and then turned around to come back. Then came the problem—she was getting tired and it seemed quite a distance to go. She settled down and rested on the netting for a few minutes—out there it's attached to the "rain" pipes. By the time she got back I was feeling better and took a short flight myself. I didn't try to go all the way to the "South Pole," though. Somebody told us that in Island Three there's going to be a cocktail bar hanging in space at the .05-g level, half a mile out from the end. That flight will get people's thirst up!

Feb. 2: Your Dad is starting to get deep in his work, as he always does, so the vacation is over for a while and I'd better take up the letter-writing. It was really thoughtful of you to make such nice preparations for our weekly call. The children looked great. I think they're getting used to the calls and aren't so shy any more.

At the docks on our way over to Island Two we met a rather sad group—some of the people from our training class, who've decided that they just can't take it here, and are going back to Earth. It's not a physical problem, because almost no one feels any dizziness in a habitat as big as Island Two. I guess that for those people all the newness and differentness are just too much, and they haven't been able to get over it. The old-timers tell us it's a well-known thing, and they call it the "wide syndrome." We didn't know what they meant until someone tipped us off: it's WAIDH, for "What am I doing here?"

After Island One, "Two" seemed really big. The basic layout and the landscaping are similar, though, except that "Two" is not so warm, and runs with a climate that's right for pine trees and firs. You know I like rhododendrons, and our apartment has a lovely mass of them against the garden wall. I don't think that Dad told you much about our apartment, but you could see some of it from the video. They've done a good thing here: because the Sun is overhead, almost, they've been able to fix it up so that there's a space of a foot or so between the apartments, and the sunshine comes down that and shines on a planter that's just outside a long window of the living room. It makes the living room sunny, and gives us very good soundproofing—we can't hear the neighbors at all. The birds are fairly noisy, though, especially in the morning just after the rain, when the butterflies first come out.

We sample the restaurants quite often, and meet people that way. They're a nice sort, and we feel very safe and secure here. Maybe because we all arrived with so little, and because most of our salaries are being put in the bank, no one seems to lock his doors. I like getting my groceries at the supermarket; that you'd really find different! The vegetables and fruit are spectacular, especially the tropical ones. At first I felt like buying up strawberries and guavas, but we're getting used to the fact that they're "in season" all the time here. Dad misses his steaks, but I tell him that back on Earth we could hardly ever afford them anyway, so he shouldn't complain. I've joined a cooking club, and the Garden Club, and am trying to duplicate a recipe we had in one of the restaurants last week. It's chicken, but cooked almost with a taste of seafood, like lobster. Next I'm going to cook a glazed ham, because with two people we can get a lot out of that for a week or more.

We both like the low-gravity swimming, especially the

diving. The water comes up to meet you so slowly that you have plenty of time to do two or three flips before you meet it:

One thing we both really like is the six-day week with only four working days. I say only four, but really there are so many clubs and volunteer jobs that we find ourselves working harder on the weekends than at our plant. Of course, it's set up this way so that the parks, the restaurants, the churches, and all the rest of the facilities get used efficiently and without crowding. With only one third of the population having a weekend at any one time, you don't find the parks empty one day and crowded the next.

Feb. 15: I could never get Dad to a ballet back home, but the Russian company from one of their communities was here last week, and we both had to see them. It was in one-tenth gravity, of course, and we both realized that ballet was really meant to be done that way. I don't know all that much about it myself, but anyone could see that all the easiness, and lightness, and the whole dreamlike quality of ballet is just so much better without gravity pulling down every motion. We came away just stunned.

I'm addressing this one just to you, Dear, because much as I love my son-in-law there are some topics I'd feel shy about with him reading the letter. All I can say is, I hope that you and he get a chance to come up here one day. We'd heard a few remarks about the zero-gravity hotel, of course, but nothing we'd heard could have prepared us for what we found. Dad had it all arranged for our anniversary, but kept it a secret from me. First he took me to a really wonderful little Italian restaurant in one of the villages high on a hill: all candlelight and soft music, a terrace with a view, and good food. Then, with only a brief stop at home to pick up our things, we were off to the Floating Island Hotel for our weekend. Most of the hotel, like the lobby and restaurants—and the showers—are at one-tenth gravity, but those bedrooms! My dear, it's just indescribable. Of course, you could watch TV or listen to music if you want, but really, as Dad says, those rooms are designed for just one thing. I can't imagine you two ever not getting along well together, but if you ever have a problem, before it gets too serious bring him up here for a

214

second honeymoon! You may never want to go back. Now that we've found what it's like, I can tell you it's going to be a lot harder for us to leave!

> *With much love–*
> *Contentedly,*
> *Jennie*

The days in which Edward and Jennie voyage to their New World are taken to be twelve to fifteen years after the completion of Island One. On the fastest possible time scale, there might be at that time a rise of the total population in space from 500,000 to one million within a two-year period. That's about seven hundred people each day; not much compared with the traffic through one of our major airports, but more than the space shuttle could cope with unless the fleet and the launch facilities were to be expanded greatly. I assume, in keeping with studies already carried out by NASA and its contractors, that well before the end of this century there will be shuttle vehicles, propelled by chemical rockets of a somewhat more sophisticated type than those of today, capable of lifting off Earth and accelerating to orbital speed without staging (dropping components) at all. Such single-stage-to-orbit vehicles are said to be within 1980s technology, so I don't think it's being rash to assume they'll exist by the late '90s or the early years of the next century. They'd bring the cost of Earth-to-orbit transfer down considerably. There is an often-quoted observation by Theodore Taylor that's quite relevant to the question of present-day space-transport systems: [1]

(Paraphrasing): "Present costs for putting freight into orbit are high for the same reasons that jet travel on Earth would be expensive if the corresponding rules were followed for the operation:

1. There shall be no more than one flight per month.
2. The airplane shall be thrown away after each flight.
3. The entire costs of the international airports at both ends of the flights shall be covered by the freight charges."

215

The way to obtain lower costs for lifting freight into orbit is evident from this quotation: develop vehicles which are fully reusable, and find a market large enough to justify frequent flights. There are, though, two "catches" in this reasoning: first, the studies which have been made so far indicate that with chemical rockets it would be extraordinarily difficult, if not impossible, to build a fully reusable vehicle capable of making a round trip from Earth to L5 without refueling. Second, the development costs for any vehicle that requires a big leap beyond the existing state of the art are very high. For the so-called "super-shuttle," for example, a vehicle capable of taking enormous payloads to orbit and of making round trips without discarding any of its components, I have seen NASA-estimated development costs of $40 to $60 billion. The vehicle which I imagine in the letters of Edward and Jennie is of a more modest sort, carrying a much smaller payload.

During the time period about which I am now speculating, I am assuming that it will not yet be practical to obtain carbon, nitrogen, and hydrogen from the asteroids. For conservatism, then, it seems safest to assume that it will be necessary to bring up from Earth about one ton of those elements for each emigrant. Such freight would not need to travel on the same very safe vehicle that would be needed for human transport.

For transport of seven hundred people per day, by single-stage rockets with payloads only two or three times that of the existing shuttle, there would only need to be about five flights each day. For a completely reusable vehicle that doesn't require assembly, only to be fueled up before each flight, such a liftoff rate doesn't seem high, even if by then we do not have additional launch sites beyond the existing two (those of the United States and of Russia). Freight requirements, though, might be higher in terms of tonnage; probably not in terms of flights. A flight every three hours or so by a shuttle-derived HLV would be enough to bring up the required supplies to initiate agriculture and to establish a comfortable environment even during the period of rapid buildup of population at L5. By the time we need that sort of freight-hauling, though, we'll probably use the same single-stage-to-orbit vehicle to do the job; a few flights each day will be enough, and that vehicle will probably burn much cleaner fuels than does the existing shuttle.

The problem of travel beyond low orbit is quite a different one: the advantages of full-time solar energy and easy access to lunar materials can only be enjoyed at escape distance, but to go from low orbit to a great distance requires a far longer time and, if Earth is still the source of supplies, a longer and thinner supply line. The problem is analogous to that of an extremely long-range aircraft flight. If we require that the plane reach its destination, turn around, and return without refueling, we make the problem far more difficult than if we permit refueling at the destination for the return trip.

The problem of low orbit to L5 transfer is, for passengers, first that of time: even with high-thrust engines, able to make large changes in the velocity of the rocket within a period of only an hour or less, the travel time to escape distance is about three days. The simple type of accommodations that would be adequate for a flight of half an hour or even of several hours would be quite unbearable for a trip lasting for a number of days. "Steerage to the stars" is not the image that we would like to look forward to in connection with the humanization of space.

Fortunately, there are compensating advantages of which use can be made to get around this problem: from low-orbital distance out, there is no requirement that vehicle engines be capable of supplying a thrust greater than the vehicle weight. If we are willing to settle for a slow trip, engine thrust and acceleration can be quite low. If we make use of the fact that L5 will be a site at which reaction mass will be relatively cheap, it seems clear that instead of developing monster vehicles for liftoff from Earth, we would be better advised to solve the problem from both ends. L5 is the ideal site for construction of large spaceships, whose design could be free of any of the limitations forced by entry into planetary atmospheres. Mass-driver engines for those ships can "fuel up" at L5 with reaction mass either in the form of industrial slag or of liquid oxygen.

The spaceships *Konstantin Tsiolkowsky* and *Robert H. Goddard* are assumed to have empty masses of about 3,000 tons, of which about two-thirds would be their mass-driver engines and their solar-power plants. The mass-driver engines would have exhaust velocities about twice as high as for the best chemical rocket—about the same as for the much earlier but similar machines studied intensively in the late 1970s for the early days of space manufacturing. Those engines, carrying solar-cell arrays like the sails on a square-rigger, would stretch out for several kilometers, but that would be quite tolerable for vessels never intended to enter an atmosphere.

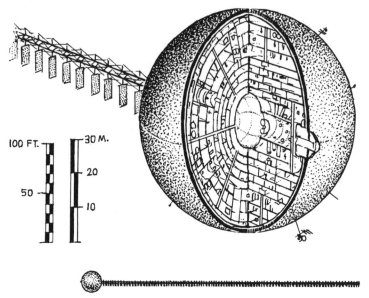

Deck layout of *R. H. Goddard*.
Rotation gives Earth-normal gravity
in lowest levels.

To find the performance of the *Goddard* we have to know how much the solar-cell arrays will weigh. I'm assuming three and a half tons per megawatt. The NASA Johnson Space Center, in a detailed study, concluded it could do that well even by the 1980s, for a satellite power station.

218

For the *Goddard*, years later in time, that should be attainable: especially so when one remembers that for a spaceship engine there is no need to hold the cost down to the low value that would be required for an economical central power station. For the *Tsiolkowsky*, the *Goddard*, and their sister vessels the corresponding travel times would be around three weeks for the inbound leg of the journey, and just over a week for the outbound: about the same time that it takes to cross the Atlantic on a medium-size vessel. The differences in trip time arise from the fact that the engine would have constant thrust and that on departure from L5 each ship would be heavy with reaction mass. That difference would be a happy one for the outbound travelers, who would enjoy a higher average speed than would the crew when spiraling down to low orbit from L5. Later on by perhaps two decades, when transport requirements may be much greater, the engineers may be able to make still lighter solar-cell arrays. If they can produce something in the ton-per-megawatt range, the travel time can be reduced to little more than three days. Other approaches, including the possibility of laser or micro-wave beamed power, are not out of the question. I am not considering the possibility of nuclear power. The reason is straightforward: if the development of the communities is to go on without check for a long period, one must not design into it "absurdities" that would pose a limit as soon as total numbers or total required transport exceeded some modest value. It does not seem to me to make sense to design a deep-space transport system around an energy source that would have to come from Earth.

We can get lower and upper limits to the ticket price for a trip to L5 in the late 1990s-early 2000s time period. The lower limit comes by assuming round trip times of a month, and ship costs per ton that are three times as high as those of present-day commercial aircraft. The total comes out around $6,000. The cost of reaction mass would only be a small fraction of that total, because it would be so abundant at L5. A still lower ticket-price could exist if the ships carry full loads of either passengers or cargo both on the inbound and outbound legs of the journey.

The upper limit is $30,000, and comes by assuming that each vessel must collect in revenue an amount equal to its own cost, within a time of eighteen months. Ticket costs on commercial jets within the United States have about that ratio to aircraft-purchase price; they include, though, total fuel costs which are a higher fraction of the cost of the capital equipment. Either the $6,000 or the $30,000 figure would be a small fraction of the productivity of an industrial worker in a single year, at the favored location of L5, and would probably equal only a few months' earnings.

Edward, Jennie, and their co-emigrants are assumed to arrive on the scene after the first settlers of Lagrangia. The first arrivals will have faced more difficult conditions and a more limited environment, but one much less harsh than our ancestors faced when the New World was opened to colonization. There will be no "hostile Indians" and there will be plenty of food.

As noted earlier, agriculture at the communities will be intensive and highly mechanized. By the scheduling of opening and closing thin shades several kilometers distant in the Sunward direction there will be long summer days in the agricultural areas, with never a cloud or storm, all year. Monotonous, but a field of grain doesn't demand variety. Temperatures in the agricultural areas will be kept always very warm, so that corn, sweet potatoes, sorghum, and other fast-growing crops can grow from seed to as many as four harvests in a year. [2] Cuttings and some grains will be eaten by chickens and pigs and turkeys, so that the settlers can enjoy a varied diet including high-protein meats. [3] There will be no need for insecticides or pesticides in the Islands, because agriculture will begin with carefully inspected seeds from Earth. The initial lunar soil will be sterile, and nothing will be introduced into it but water, chemical fertilizer, and necessary strains of bacteria helpful to plant growth.

Though beef cattle may be too wasteful of space and too inefficient in converting plants to meat to earn space in the early communities, there will be good reason to include a small stock of dairy cows: children will still need their milk.

In the villages of the Islands there must be insects,

perhaps butterflies, for the birds to eat, but there need not be mosquitoes—or cockroaches, or rats. The space clothes described by Edward and Jennie can be light, because their environment will be free of harsh extremes. They will find no "wide open spaces," but then neither do the inhabitants of northern areas on Earth, who must pass long winters mainly indoors.

In trying to guess what life will be like for the early settlers, we should recall one of the most deep-rooted of our own needs as healthy human beings: the need to feel that we are of value, that our effort and work are needed and appreciated. In Island One everyone will have the sense that his work is needed and is important; there will be no unemployment. Probably the early residents will develop close-knit communities, and their villages may develop identities of their own even though they are no more than a few minutes' travel-time apart.

Many of the enjoyments of the early communities will be those which we would expect in a small, wealthy resort community on Earth: good restaurants, cinemas, libraries, perhaps small discotheques. Yet some things will be very different; there will be no cars and no smog: travel will be on foot or on bicycles. By the time the first Island Two is built, it may be possible to have not just the small river imagined for Island One, but a lake; with inexhaustible solar energy close at hand that lake may have beaches lapped by waves perhaps even large enough for surfing.

11
HOME-STEADING THE ASTEROIDS

Our ability to send people far into space, and there to maintain them in good health, reached its limits with the Apollo Project. The life-support systems developed for that venture were capable of maintaining human life for two weeks, long enough only for a quick dash to the Moon, a few days of exploration, and a return. The Skylab Project of the early 1970s extended the time limit for astronauts in space to three months, but in a location much closer to the earth, in low orbit. [1]

The maintenance of life for a time of many months, as would be necessary for a voyage to the asteroids, presents no problems which are new in principle, but the detailed engineering of such systems has not been done. We should assume, then, that the early space communities will be built entirely of materials from Earth and the Moon. As the number and size of the islands in space increases, there will be a demand for additional materials to stock them. That demand will be a strong incentive to tap the resources of the asteroids for carbon, nitrogen, and hydrogen, though it is likely that for some time all other elements will be obtained from the lunar surface.

A serious proposal for the humanization of space could not have been made before the Apollo lunar samples were returned to the Earth. Similarly, it is only in the past several years that our information on the compositions of the asteroids has advanced from the stage of speculation to that of near certainty. The recent increase in knowledge and understanding has come from the development of three new techniques [2,3]: the measurement with high resolution of the dependence on wavelength of the reflection from an asteroid of sunlight, in the visible and near-infrared region (spectrophotometry); measurement of the polarization of the sunlight reflected from an asteroid (polarimetry); and the measurement of infrared light from an asteroid (radiometry). The last two methods combine to give a measure of the diameter and the average coloring (light or dark) of each asteroid, and the first is now so sophisticated that spectra characteristic of particular minerals can be recognized in the light reflected from an individual asteroid.

More than 90 percent of the asteroids fall into the classifications "carbonaceous chondritic" or "stony-iron"; these classes correspond to groups of meteorites found on the surface of Earth. Carbonaceous material is not unlike oil shale, being rich in hydrogen, carbon, and nitrogen. It is generally soft and friable, and can be melted at a low temperature. Probably for that reason, not many car-

224

bonaceous meteoroids survive their fiery passage through our atmosphere. In the more benign environment of the asteroid belt, much more of the carbonaceous material has survived; there is fairly good evidence that most asteroids are carbonaceous, including the two largest of those minor planets, Ceres and Pallas, with diameters respectively of almost a third and about a seventh that of the Moon.

The energy interval between the asteroids and L5 is almost exactly the same as that from the earth to L5. For a practical rocketeer, that energy interval is expressed by velocity changes that must be made in order to change the orbital radius and to tilt the plane of the orbit from that which matches an asteroid to that of Earth and Moon. Asteroids in the "main belt," out beyond Mars, move relatively slowly in their orbits. Earth, nearer to the Sun and therefore more strongly attracted to it, must travel faster in order not to be pulled deeper into the Sun's range of gravitational influence. The difference is typically six kilometers per second, and must be made up in the course of any voyage to or from an asteroid. Further velocity changes must be made to match the eccentricity (lack of circularity) of an asteroidal orbit. Most asteroids circulate in planes inclined to that of Earth (ours is called the "plane of the ecliptic"). For each two degrees in angle by which the planes differ, an additional velocity change of about one kilometer per second must be made. If one searches through the list of asteroids for those with favorable orbits, and calculates the total velocity interval which separates each from L5, the answer in nearly all cases turns out to be near ten kilometers per second. The velocity interval between Earth's surface and L5 is only slightly higher.

Although these velocity intervals to L5 from Earth and from the asteroids are so nearly alike, there will be at least two incentives for obtaining carbon, nitrogen, and hydrogen from the greater distance rather than the lesser. In deep space, high thrusts will not be required, nor will spacecraft hulls to protect payloads during their brief passage through Earth's atmosphere. In the long run, the economies made possible by those additional freedoms will almost surely tip the transport cost scale in favor of the asteroidal resources. That shift, when it occurs, will

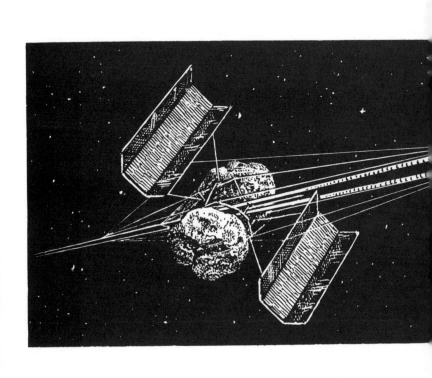

covery of asteroidal
nks by twin-engine
ss-driver tug.

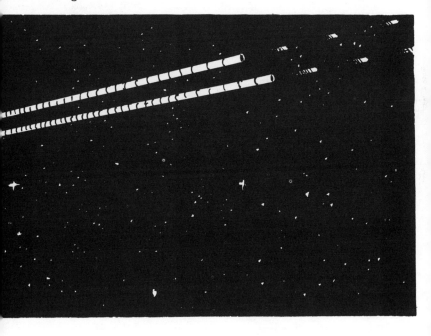

avert what would otherwise become an increasing burden on the biosphere of Earth from rocket flights through the atmosphere. Eventually, it seems likely that transport up to low orbit, and from there to L5, will be needed only for people and for particular products, especially those which are light in weight, but which can only be made by specialists from among Earth's large population.

The time delay before the asteroidal materials can be exploited may be estimated by the duration of the Apollo Project. In that case about eight years was required to progress from the early, primitive earth-orbital flights to the successful round trips to the Moon, a thousand times farther away. In order to go to the asteroids, it will be economically advisable first of all to have a well-established facility at L5. The early space manufacturing communities can supply as by-products of the industries reaction mass for impulse engines. Those communities can also serve as shipyards, for the vessels of deep space. By contrast, an asteroidal journey starting from Earth would require several times as much energy as from L5, and would involve the added expense of vehicles for precision lift-off from the planetary surface under conditions of strong gravity. If the construction of an asteroid-voyaging ship is begun several years after the first L5 community is established, the first human venture to the asteroids might begin within eight years from the dedication of Island One. It would be preceded by relatively inexpensive unmanned probes, also launched from L5, so that the first travelers would go to an asteroid which is already known to contain the elements wanted. The situation is quite unlike that of speculative oil-well drilling into the surface of Earth; we can know more about the composition of an asteroid a hundred million miles away than we can know, without drilling, about Earth a thousand meters beneath our feet. In space there need be no wildcat oil operations or dry holes.

For ecomomy, our transport system should be the analog of Earth's cheapest: a tugboat and a string of barges. In space, where there is no drag, we could practice a further economy: the tugboat would be needed only at the start and end of each trip, and during the long months of the orbit from the asteroidal source to the region of L5 the payload, in the form of tanks of ammonia and of hydrocarbons, could travel unmanned.

Like Earthbound tugboats, ours would be mainly engine and not very beautiful. One conservative design could be based on a longer version of the mass-driver used on the Moon; it could be many kilometers long, if braced by yardarms and wires like the mast of a racing sailboat. The structure could be lighter if the payload were distributed at intervals along the whole length of the engine, because the thrust of the engine would be distributed equally along its length. The tugboat might be powered by lightweight photovoltaic solar cells, assisted by large, lightweight mirrors to concentrate the weak sunlight of a distant orbit. The active electrical components of the mass-driver might be contained in a long, thin aluminum tube, pressurized with oxygen to the equivalent of a mountain altitude on Earth; that would permit maintenance and repair of all compenents likely to give trouble, without the inconvenience and lost efficiency that accompanies the use of space suits.

There would probably be living quarters for six or eight people, sufficient for three watches as on a boat at sea. There would be a small chemical-processing plant, sufficient to form reaction mass from asteroidal debris. Altogether, the tugboat might have a mass of a few thousand tons, comparable to that of a large Coast Guard icebreaker on the oceans of Earth. The payload, in the form of tanks of chemicals, could be as much as the cargo of an oil tanker. After months of steady pushing the payload would have acquired the velocity changes necessary to put it on orbit to L5. The tug would then cast off, and return home to the asteroidal outpost-community. The returning crew would take time off while the tugboat was piloted on its next trip by a rested crew. Meanwhile the linked payload would swing silently inward toward the sun on an eight-month flight that would bring it to the vicinity of L5, and rendezvous with another tugboat for the final velocity change.

Tugboats on the oceans of Earth, even exposed as they are to storm and damage, often last fifty years. It has been

Mining an asteroid for reaction mass.

one of the phenomena of the early years of experience in space that satellites usually last much longer than their "design" lifetimes. For the mass-driver-powered tugboats of the asteroidal belt, which would operate without high temperatures or pressures and would never be exposed to wind or storm, the lifetime would probably be much longer. Probably they would be retired by obsolescence rather than by wear. Transport costs for material from the asteroid belt, based on present-day figures for interest rate, amortization, and costs of aerospace equipment, lie in the range of less than a dollar to several dollars per kilogram. That's far higher than the cost of supertanker transport on Earth, but much lower than the costs for any presently conceivable transport system operating from Earth's surface to L5.

As has happened so often when we've studied in depth possibilities that seemed promising as aids to space manufacturing, the asteroids may be even better sources of materials than I've suggested so far. Though most of the minor planets are in the main belt, Dr. Brian O'Leary pointed out that a special class, named after the asteroids Apollo and Amor, have orbits much closer to the Earth's. In a 1977 NASA-Ames study O'Leary gathered leading experts on asteroidal measurement and orbit theory. They worked out detailed scenarios for recovery of specific, known asteroids of the Apollo-Amor class, using mass-driver reaction engines. Their technique made use of "gravity assists," and in action that would be spectacular indeed: after rendezvous with an asteroid, the tugboat crew would so direct their engine that the asteroid would swing by a planet like Venus or Earth. At the swingby, the asteroid's velocity would be changed as much by the planet's gravity as it would by months or years of mass-driver operation.

With the help of the gravity-assist technique, already well-proven in spaceprobe missions to the outer planets, it seems that some of the asteroids may be much more accessible than those of the main belt, and from an economic viewpoint may even give the Moon a run for its money. There's plenty of material available; even the smallest asteroid we can see in our telescopes has a mass of more than a million tons.

At a certain point in the growth of the L5 communities, trade between the islands of space will begin to dominate over the "colonial" economy of interchange with Earth. We have seen that transition take place in the colonies of the Americas, Africa, and Australia. It seems likely that for any new community whose major purpose is the habitation and maintenance of its population, rather than of supply to L5 or to the Earth, economics will favor its construction without any prior shipment of materials at all: that is, in the asteroid belt itself.

The construction equipment for building a new habitat could be sent to the asteroids from L5, or manufactured in the asteroidal region. With that equipment new habitats could be built from material readily at hand, and as soon as each new habitat is ready its population could travel from Earth or L5 to occupy it. The saving in transport cost by that development would be significant; the weight of the settlers who would move into a new habitat would be only about one five-thousandth of the weight of the habitat itself. Again, there's plenty of material; to build a colony the size of Island Two, for a population of more than a hundred thousand, an asteroidal chunk a few city blocks across would be sufficient—a mere speck, at the margin of visibility from Earth.

Once in operation, a space community would be quite capable of moving itself, in a leisurely fashion, to another point in the solar system. To do so in a manner economical of reaction mass might require a technology presently being studied, but not yet realized at the level of engineering practice: that is the acceleration of tiny pellets or grains of solid material by electrostatic forces. The ion-rocket engine, a device already built and tested in the anticipation of scientific probes to the asteroids, works by the same principle; the difference lies only in the size of the pellet being accelerated. The ion engine would accelerate something more nearly like a grain of dust.

Until the intensive theoretical study of mass-drivers in the late 1970s, they would not have been thought of as serious competitors to ion thrusters for high-performance missions. Now, though, it appears that a mass-driver might perform quite well even in the demanding assignment of moving a completed habitat through some great distance within the solar system.

During the development of chemical rocket engines, exhaust velocities have increased steadily. The higher the velocity of the exhaust, the less fuel need be carried for a given task. In the case of an ion or pellet engine, though, high velocity is not always desirable. The velocity of an ion, in the case of engine with parameters that permit easy operation, is so high that the performance is limited by the electric power available. If one halves the exhaust velocity of an ion engine, the reaction mass required to carry out the mission in the same length of time doubles, but the electric power required decreases to half of its previous value. For any given task there is an optimum exhaust velocity, just high enough so that the expenditure of reaction mass is not intolerable, but low enough to minimize the amount of electric power needed for the engine.

In the case of a moving island in space, the optimum exhaust velocity is five to ten times that of a chemical rocket, if the task is to move the newborn community from the asteroid belt to the vicinity of L5. For such a velocity the amount of reaction mass used in the trip would be only a quarter of the habitat mass. It would be obtained in the course of the voyage by processing a cargo of asteroidal rubble, possibly by a simple grinding and seiving operation. The lifetime of a community would be indefinitely long, given continuous habitation and maintenance; on a time scale of at least thousands of years, it would not seem unreasonable to devote thirty years to a relocation. Based on present-day costs for turbogenerators,[5] the necessary power-supply installation for that task would be capitalized at $25,000 to $60,000 per inhabitant, certainly not an exorbitant figure. In the last chapter of this volume I will describe just how far a community could go, if possessed by wanderlust. For the moment, though, it is enough to point out that the choice of location might be made by a vote of the inhabitants, and that the choice might not always be that of returning toward L5. Any orbit within the entire volume of the solar system, out to a distance farther than that of Pluto, could be reached within less than seventy-five years by a space community; within that huge volume it would always be possible to obtain a full earth-normal amount of solar intensity, by the addition of lightweight concentrating mirrors to the light-reflection system that an ordinary habitat would carry. A community or a group of communities desiring a peaceful and quiet life might well choose not to return toward Earth, but to "go the other way" to a private orbit from which the interaction with the population near the Earth would be, at most, by electronic communication.

We should realize that the humanization of space is quite contrary in spirit to any of the classical Utopian concepts. At the heart of each Utopian scheme, including the modern communes, there have nearly always been two very different, even conflicting ideas: escape from outside interference, and tight discipline within the community; freedom and constraint.

Escape from outside interference will be an option open to a community in space, unless military intervention occurs to prevent it: there will always be the possibility of "pulling up stakes" and moving the habitat to a new orbit far from the source of the interference. In history we have many examples of groups, not least among them our Pilgrim ancestors, who have been permitted to escape from coercive situations. Usually those who remain behind justify that permission by something equivalent to "We're better off without those troublemakers." The space communities would be in contrast to the classical Utopias in part because they could escape so much more successfully. Here on Earth the possibilities for escape are limited, because a community that desires isolation is still forced by climate and the scale of distance to become part of a distribution system thousands of miles in extent. Indeed, one of the unpleasant characteristics of modern industrial life is that regional differences tend to be ironed flat by the economic pressures toward uniformity. The differences between small villages in separate countries are now far less than they were a generation ago, and something has been lost in that transition.

The communal enclaves of nineteenth-century America, the Shakers, the Mennonites, the Pennsylvania Dutch, the Oneida Community, and others, nearly all consisted of groups each of which was united by an unvarying, agreed-on plan for how people should run their lives. Those who have lived in and then left the modern communes tell us that however the codes of behavior of these organizations may differ from the norm of the world outside them, internally they have rules strictly maintained. This should be no surprise; a commune is the limiting form of a small, isolated village, and as anyone who has lived in such a place can testify, social pressure there is almost always far stronger than in the anonymity of a large city.

In contrast, and very much by intent, I have said nothing about the government of space communities. There is a good reason for that: I have no desire to influence or direct in any way, even if I could, the social organization and the details of life in the communities. I have no prescription for social organization or governance, and would find it abhorrent to presume to define one. In my opinion there can be no "revealed truth" about social organization; there can only be, in any healthy situation, the options of diversity and of experimentation. Among the space communities almost surely there will be some in which restrictive governments will attempt to enforce isolation, just as such governments do on Earth. Others, hopefully the majority, will permit travel and communication. Within the brief time of twenty years, during which transatlantic air travel has gone from the unusual to the commonplace, we have seen how powerful a lever it has been for the transmission of experience from one country to another, especially among the fraternity of young people. Logically, if the cost of transportation between the communities becomes as low as it is now projected to be, travel between most of the communities of space will be far more frequent than it is now between nations on Earth, and people will be able to form their own opinions, on the basis of direct observation, as to what constitute successful or unsuccessful experiments in government. With energy free to all, materials available in great abundance, and mobility throughout the solar system available to an individual community, it should be more difficult in space than it is on earth for an unsuccessful government to argue that its failure is due to unavoidable circumstances of location or resources.

There is another profound difference between the historical Utopian attempts and the humanization of space. The

235

communities of the past were formed on the basis of new social constructs, but took their technology from the world around them. Some even made a conscious selection of more primitive or more restricted technological equipment than available in the world outside. In extreme form this tendency shows in the prohibition, among several of the existing Utopian sects, of any equipment for day-to-day living more advanced than that which was available in the nineteenth century.

The reason for this restriction, usually clearly stated and understood, is the need to prevent "contamination" of the Utopian social ethic by contact with the outside world. There is recognition by the leaders of the enclave that its social organization is unstable, and can only be maintained by isolation. Usually, the "danger" to the maintenance of that unstable situation is that young people from within the enclave will learn of the additional choices available in the world outside, and will insist on leaving to enjoy them.

I share with many an admiration for the Utopian groups that have managed to retain their identity and values through several generations of rapid change. Those of us who might have been tempted, during the decade of the 1950s, to feel concern and even sorrow because of the narrowed horizons permitted to the children of such groups surely felt quite differently during the 1960s, seeing an epidemic of drugs and a lack of purpose spread throughout a generation in the world outside. It may even be that among the existing Utopian groups there are some free of antitechnological taboos, which will find it easier to retain identity by resettlement in space than to remain on Earth. The humanization of space is though no Utopian scheme: the contrast is between rigid social ideas and restricted technology, on the part of the Utopias and communes, and the opening of new social possibilities to be determined by the inhabitants, with the help of a basically new technical methodology, on the part of the space communities.

One can speculate, with some supporting evidence, that as a result of the individual choices which led to the historical colonization movements on Earth, there are now subtle but real differences in attitude toward change and further migration on the part of the people in the old and the new countries. Here in the United States, and in Canada, Alaska, Australia, and other former colonies, there may be a greater restlessness, a greater desire for travel and change, than exists in those populations descended from the people who stayed at home rather than emigrate. Of the thousands of letters I have received about the space-community concept, a disproportionate number come from the lands that were once colonies. Already, from the many letters that express a personal desire on the part of the writers not just to support but to take part in the outward venture, it is clear that the early settlers in space will be exciting people: restless, inquiring, independent; quite possibly more hard-driving and possessed by more "creative discontent" than their kin in the Old World.

In space, where free solar energy and optimum farming conditions will be available to every community, no matter how small, it will be possible for special-interest groups to "do their own thing" and build small worlds of their own, independent of the rest of the human population. We can imagine a community of as few as some hundreds of people, sharing a passion for a novel system of government, or for music or for one of the visual areas, or for a less esoteric interest: nudism, water sports, or skiing. Of the serious experiments in society-building, some will surely be failures. Others, though, may succeed, and those independent social laboratories may teach us more about how people can best live together than we can ever learn on Earth, where high technology must go hand-in-hand with the rigidity of large-scale human groupings.

Just as happened during the settlement of the American West and of Alaska, when the population at L5 increases in number some of the pioneers may be the sort of people who will say: "It's getting too crowded around here; let's move on." Those people may be among the most interesting and productive individuals. They may want a more complete independence, and so may decide to go homesteading just as did our great-grandparents in the mid-nineteenth-century American plains states.

Homebuilt spacecraft carries
family on homesteading voyage
to the asteroids.

Here, now, is one way in which a pioneer family might go about a homesteading venture. Though the details will surely be different from those I describe, each possibility that I will give is based on a number that can be calculated, or on analogy to similar situations here on Earth. I am giving it in the form of excerpts from a diary, written perhaps in the early years of the next century. That too is by analogy; one of the relics of my family, preserved through five generations, is a book by an old lady who must have been, in her time, a holy terror. In her eighties she wrote an account in verse of a time when she had traveled with her seven sons across the plains of America in a covered wagon. In their travels the little band encountered dangers that space settlers will not face; hostile Indians, snows, exposure, and short rations.

July 15, 20–: Dear Stephen:

Your Mom and I are going to write down a record of our trip, to go with the pictures we're taking. Then when you're old enough to read and be interested in it you'll be able to see how you came to be a youngster living in the asteroid belt.

238

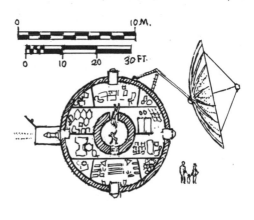

It's been five years, now, since I joined the Experimental Spacecraft Association. We have an active chapter of it here on Bernal Gamma, and several of the guys in it work with me in the construction business.

If we were back on the Earth, now, and got any wild ideas about setting out on our own to travel in space, we'd be out of our minds. A spaceship that could lift its own weight, and go through the split-second timing that you need for a lift-off from Earth, would be a lot more complicated and expensive than any home craftsman could build.

Out here, though, we're in much better shape to go voyaging on our own. Our spacecraft never has to take big forces, and our engine can be small; we don't mind taking quite a while to get somewhere.

With what we'd saved, and the sale of our house on Gamma, we were able to start with about $100,000. For the past three years I've been working on the spacecraft, and we'll hang on to it when we arrive in the asteroids, so it'll still be around when you're old enough to remember things. The *Lucky Lady* is a sphere about three stories high, made of aluminum because that's easy to weld. I've been building it in the marina, near the docking ports of Gamma, and we've checked the welds with x-ray equipment that we've borrowed from the plant. Alongside the *Lady*, at the marina, there are four more of the same kind; Chuck and Bill and the others will be going with us, in a "wagon train" of five craft, so that if any of us runs into trouble either before or after we arrive, there'll be help near at hand. Between us five we've bought a complete spare engine and a lot of spare parts and one-of-a-kind tools. When we get to the asteroid belt we can team up for big jobs when we have to.

Our plans came out of *Spacecraft and Pilot*, and were checked over by astronautical engineers before they were published, so they're sound. The *Lady* has a triple-pressure hull, each layer a millimeter thick. and any one of the layers would be enough to hold a lot more pressure than we'll need. Altogether the bare hull weighs about 3 tons, and there's a lot of my time that's gone into it. The marina doesn't rotate, so all the construction was done in zero-gravity. That way, I could handle the big sections of aluminum by myself.

Around the hull there's a layer of sand about a foot thick, to protect us from some of the cosmic rays and from solar flares. Outside the sand is a fourth shell, of very thin aluminum just to hold the sand in place. For extra help in case of flares we've also got a "storm shelter" outside the sphere in the form of a small aluminum bubble connected to the big one. There, the shield is a lot thicker, and if a flare starts we can be in the storm shelter in less than a minute and can stay there for several days if we have to. Babies are extra sensitive to cosmic rays, so the "storm shelter" is your nursery too.

We bought our rocket motors new. They're from the same company that makes them for the small Coast Guard rescue boats. Each one gives a thrust about as much as my own weight, and a bigger chunk of our "grubstake" money went into those than anything else. I understand they cost about the same as a small jet engine on Earth. Our life-support air-recycling system was bought used, rebuilt and recertified by the Federation Astronautical Agency. It too came off one of the Coast Guard boats, and we got it cheap, but I know that the government paid a lot more. They've gone over to newer models now.

Back on Earth, before your Mom and I moved out here, I used to belong to an Aero Club and flew little airplanes for fun. Things happened fast, there, and navigation in bad weather kept me on my toes; I'd have to keep track of the Omni-Range signals, and the direction finder sometimes, and stay legal as far as the control altitudes and the rest of the regulations went, and all the time fly the airplane by reference to the compass and the gyroes. Going from here out to the asteroids I won't have to worry about all that; there's no weather in space, so we'll be able to see where we are and where we're going all the time. We'll have two systems for navigation. One of them is as old as sailing ships on Earth's oceans: it's a sextant, to measure the angles between the visible planets and the Sun. That would be enough to do the job, but we've also got something else. Nowadays there are big transmitters set up in the orbits of Earth and Mars, and they send out pulses so we can calculate our position just by a simple radio receiver. On Earth's oceans they used the same method for navigation, and called it Loran. With the handbook of transmitter positions and times that we've got, we can figure our position to within less than a mile, even though we may be twenty million miles out.

We went a bit overboard on radios, and bought three, all alike. They're about the size of the ones used in small airplanes. We'll use them for voice communications between the five families traveling together, and for dot-dash Morse code to check in with the Coast Guard. We're going to be on a flight plan and will have to check in once every three days. To do that, I'll be aiming the big aluminum-foil dish antenna that I've built, using a little telescope to point it exactly back to the location of the receiver at L5.

Aug. 1st, 20–: The Coast Guard and the FAA people have been aboard, and we've got our clearance. They checked our Space-worthiness Certificate (Category R, Experimental Homebuilt) and our radio licenses, and my pilot's license (Private Category, Deep Space Only, Flight Within Planetary Atmospheres Prohibited). We've got food on board for two years, if we have to stretch it, lots of seeds, fish, chickens, pigs and turkeys. To get things started when we arrive, we've sunk about half our grubstake in prefabricated spheres and cylinders, aluminized plastic for mirrors, chemicals for crop-growing, and a lot of equipment.

Aug. 8th, 20–: The Lucky Lady, loaded, shielded, and ready to go, weighed in at close to 500 tons, so we didn't take off from Gamma with any big burst of acceleration: we weren't up even to walking speed a minute after we started thrusting, but our speed slowly built up, and now after a week we've gone farther than the distance from the Moon to Earth. It'll be another eight months to go, about as long as your great-great-great-Grandad took to get from Illinois to California.

October 10, 20–: We've had a bit more excitement than we bargained for these past weeks. First of all, Bill's engine developed a problem; he wasn't getting the thrust that he should and the fuel was going too fast. Those engines are pretty complicated and we weren't able to solve the problem on our own quickly, so did an engine-change to the spare. That wasn't too difficult: we just maneuvered the five spacecraft close together, docked them, closed up the hatch behind the engine, and did the engine-change in our shirtsleeves. From now on we'll have plenty to keep us busy, because we have all the manuals on the engine and we're going to take our time and see if we can figure it out well enough to fix the one that we pulled from Bill's ship. While the engine-change was going on we were "dead in the water" with no thrust for nearly four days. but here in space that doesn't mean an emergency. We still had our speed, and all that the lost time means is that we'll make a very small change in the thrust direction and take a little longer arriving.

Only two days after we got finished with the repairs, we got hit with our first big solar flare. Those things build up in minutes, so there wasn't time to get any warning. When the alarm bells sounded we all scooted for the storm shelters, and stayed holed up for three days; by then the flare had died down so much that our ordinary shielding was enough.

Nov. 23, 20–: We brought you out of the nursery so you could be with us for our Thanksgiving dinner: turkey, canned cranberries, and lots of extras we've been saving. So far we've got a lot to be thankful for: there were some colds early in the trip, but after that everyone's been healthy, and nobody's got any tooth problems yet. If we can last to the Belt, where there are dentists, we'll have escaped the biggest problem that hits groups like ours.

All of us have been using our time to get a head start on construction. We began with our assembly bay, and that's something the five families will share, 'til we can build more. It's a cylinder as big around as the *Lucky Lady*, and as long as a city block. It's made of aluminum sheets, and we made it without ever going out in "hard suits." We're in free flight now, the engines have been shut down, so we handled the construction bay by just clamping on to it with our grippers, very slowly walking the whole ship over to the place we wanted to work on, and then handling the welding equipment through sleeves that we've built in to each ship. I guess the setup is a bit like a chemist's dry box. The ends of the bay are hemispheres of aluminum, and when the last weld was done and checked the bay was a gas-tight chamber. We let the liquid oxygen tank get a bit of sunlight, so it would slowly boil off, and after a few days the oxygen pressure in the bay was breathable. We have all five ships locked on the bay now, so any of us can go in there to work, and that's where all the glasswork is being done. The welding, of course, is better done in vacuum.

Our first "dockyard" job has been the crop-modules. Each one's a cylinder of a size that just barely fits in the assembly bay, when all the pieces are welded together. When we're done we weld in a lightweight floor, and under that we set up the chicken coops and the pig pens. The roof is trickier, because we have to let in the sunlight. In the L5 communities they do that with thick metal webbing and then plates of glass to form the windows, but here we do things in an easier way: we have prefab aluminum sheeting that has a lot of small holes in it, and we seal over each hole a disc of glass with a plastic compound. When we finish a crop-module we pump the oxygen from the bay into a cold-storage liquid oxygen tank, and open the end-bolts and take off one of the hemisphere-ends, and float out the finished section.

Dec. 25, 20—: You were out of the nursery again today, and all twenty-three of us got together for a real big Christ-

Formation of strong metal
sphere by vapor deposition in
vacuum and zero gravity.

mas dinner. We had ham and a lot of frozen food, but next
year, if we're lucky, we'll have fresh sweet potatoes and
corn and fresh pumpkin pie as well. I've been whittling

some new toys for you, and you seemed to go for them. I know you won't thank me for reminding you when you're a bit older, but Mom is proud that you say "Mom," and "Dad," and "ship" and "dog." I don't think Chuck's family would think of going anywhere without Snoopy, and if that other dog Maggie comes through like she looks, we're going to get one of the litter for you.

May 10, 20—: Looks as if we won't have time for any more writing for a while. We've been prospecting for the past month, and now it looks as if we've found us a good one. You couldn't even see it through a telescope from the Earth, but we figure it's got a mass of around three million tons—a lot more than we'll need even in your grandchildren's time. The little spectroscopes that we brought along tell us that it's got plenty of carbon (we picked the asteroid because it looked good and black) and there's nitrogen and hydrogen and plenty of metals too. So we've got some clearing and stump-pulling to do, and by the time you're big enough to handle a welding machine you'll be my helper. We've got a whole world to build here, Stephen, so grow up fast and get in on the construction!

The spirit of adventure, and the drive to be free and run one's own life, even at the expense of hardship, danger, and deprivation, are as old as humanity, and must have been at the heart of the Westward movement as they will be for the migrations that will start at L5. If we traced the development of an embryonic settlement, of the kind that might begin with a trek of the sort just described, we might find that the pioneers would construct their habitats by the labor-saving method of evaporation from an aluminum ingot suspended by magnetic forces in zero-gravity, and heated by concentrated sunlight. Within two or three years a sphere with a land area of more than a hundred acres for habitation, and an additional several acres for crops, could be made in this way, most of it quite possibly by a housewife monitoring a control computer from her kitchen. A computer to do that job wouldn't be much more complicated than a pocket calculator, and a few decades from now a much more powerful computer installation, of the sort that's now found only in offices and laboratories, will be of desk-top size and won't cost more than an automobile. Almost certainly each of the pioneer families will be equipped with one of them.

Examining growth rates, we find that the tiny asteroidal chunk described in the homesteader's diary would suffice for a population of at least 10,000 people, so there would be no need for the pioneer group to seek new materials for at least several hundred years, even if its population grew at the present world-average rate.

In our modern world, with its concern for vanishing resources and for preservation, our immediate reaction on hearing of an available resource is to consider its protection. When I described the resources of the asteroidal belt to a group at the National Geographic Society, there was an immediate reaction: "Please don't take Geographos!" There need be no fear of that; Geographos is a small asteroid now thought to be of the stony-iron type, and should be safe from mining activity.

In the case of a growing technological civilization, with each new material resource we must associate a time scale. For example, if the total reserves of material to be found in a new "mine" will last only ten years, but if the new technology required to exploit that resource will take twenty-five years to develop, the expected returns are hardly sufficient to justify the effort. Earlier I pointed out that the material reserves in the asteroid belt are sufficient to permit the construction of new land area totaling 3,000 times that of the earth. In making that statement my purpose was not to encourage a corresponding growth of the total human population, but rather to suggest that materials limits alone should not be used as the justification for the imposition of limits on individual human freedoms. The freedom to have as many children as a family wants is by no means as important as the freedoms of speech, communications, travel, choice of employment, and the right to an education, but it is very difficult to abrogate one freedom without compromising others. As Heilbroner has pointed out, in a society held by law to a steady-state condition, freedom of thought and of inquiry would be dangerous, and would probably be suppressed. [6]

In the same spirit, not of encouraging thoughtless growth but of opening possibilities which will encourage the extension rather than the curtailment of freedom, we can look beyond the materials limits of the asteroid belt and inquire as to the total resources of the solar system. I've argued that a growth rate about a tenth as large as our present explosive increase would be sufficient to make the difference between stasis and change; it's just enough to be noticeable over the lifetime of a single human being. In the space communities, that growth could be matched by a corresponding increase in the total land area, rather than by additional crowding, as on Earth. For that moderate rate of growth, the resources of the asteroids would be sufficient for at least four thousand years, at a population density the same as that of Earth (averaged over all the land area of our planet, including the desert, polar and wasteland areas now uninhabitable).

If we look beyond the resources of the asteroids, there are three further aggregations of materials within the solar system, each of which has a large total quantity: the moons of the outer planets, the cometary debris, and the outer planets themselves. As far as we know, all of these aggregations are without intelligent life, and all but the outer planets are invisible to us without telescopic aid.

The moons of the outer planets have a total quantity of material roughly 10,000 times that of the asteroids; the outer planets themselves, a thousand times more. The existence of those resources, beyond those of the asteroid belt, means therefore that even without the cometary material there would be enough for expansion at a moderate rate for more than twelve thousand years. Each of the new classes of material resource would permit, by its exploitation, several thousand more years of expansion, and the technology required for the opening of each resource would hardly require more than some tens of years to develop. Although I don't advocate it, I must conclude therefore that there is room for growth at a moderate rate for many thousands of years, should that be desired in every era by the human population then alive.

Although twelve thousand years is short on the time scale of evolution, it is a very long time on the scale of social institutions. If we consider a voyage in time as far into the past as we can now contemplate toward the future, we would be close to the time of the last Ice Age, long before the earliest beginnings of recorded history.

If long-term growth may indeed take place, it is tempting to consider the corresponding increase in what we might call "capability," a measure of the power of humanity over the physical environment. We can only guess, but if we take the capability to be something akin to a gross national product, we may guess that it could be proportional to the growth factor itself (that is, to the crude ratio of populations), and to the productivity (the output per individual human being of some measurable product, either material or informational). If the latter is taken to be as little as 1.5 percent per year, and the former is 0.2 percent per year, the increase in total capability over so long a time as 12,000 years would be a truly astronomical factor of ten to the eighty-eighth power. The implications of that increase in capability, admittedly speculative in the extreme, are fascinating to contemplate. Almost certainly they would include an enormous degree of control over the environment by each individual human being. Ten to the eighty-eighth power, for example, is more than the number of the individual atoms in all the stars, planets, and dust clouds of our galaxy.

Evidently, then, it is possible in principle for a civilization to advance from prehistory to a state of enormous capability on a time scale which is very short in galactic terms: less than one part in 200,000 of the age of the Sun. Why, then, has no previous "explosion" of a civilization into a situation of great physical power not left its mark on the galaxy? Why are there no beacons burning to light our way? Perhaps the birth of a civilization capable of migration into space is extraordinarily unlikely, or perhaps social instability and stagnation are overwhelmingly powerful civilization-killing forces, or perhaps—as I have suggested earlier—moderation and empathy come with technical maturity, and there do exist long-lived galactic civilizations all of which prefer for our own good to let us develop on our own.

12

THE HUMAN PROSPECT IN SPACE

Speculation about a development still in the future is a scary process from the viewpoint of a scientist. He is used to making predictions which cannot be proved or disproved until later, but he makes them on the basis of experiments, carried out with all the care and diligence that he can muster. If he has maintained a sufficiently high professional standard in his experimental technique, he knows that later work can only prove him right. When a scientist indulges in speculation, he throws away the experimental tools which give him his only claim to authority and expertise, and his predictions do not deserve much more weight than those of anyone else. Even so, inevitably I must indulge now in speculation, and I do so with considerable apprehension, knowing full well that I am edging out farther and farther on a very shaky limb. Like an automobile driver in winter on an icy road, trying always to keep at least one pair of wheels on the solid pavement, I will try to keep each speculation within the bounds of numbers which can be calculated.

History and analogy are solid ground within the treacherous marsh of speculation. We know that foreign trade has been the economic basis for most of the successful human colonies in their early stages. For the long-term economic viability of communities in space, we expect that there must be something which Earth must buy from L5, and something that the residents of L5 must import from Earth.

The need for cheap energy on the surface of Earth, in the form of electricity transmitted by microwave from solar power stations in orbit, is likely to exist for a long time. Even if per capita income in the developed world remains frozen at some level, for several more generations there will be a demand for more energy every year, as the Third World struggles to achieve economic freedom and take its place in the community of nations. While that demand continues the L5 communities should find a ready market.

The suitability of L5 as a location for the production and use of heavy scientific equipment (telescopes, research spacecraft both manned and unmanned, and laboratories for the study of zero-gravity conditions) should give the residents of L5 another sector of trade with the inhabitants of the earth.

In my view, the likelihood that marketable products can be returned profitably to the surface of Earth from L5 is much more doubtful. That return would require throwing away the single biggest economic advantage that the L5 communities will have: their location at the top of the 4,000-mile-high gravitational mountain which towers above us here on Earth. Nevertheless, some consideration should be given to this possibility. The mechanics of payload return have been considered by Eric Drexler of M. I. T. He concluded that shipments of materials from L5 to Earth might best be made within reentry bodies fabricated of titanium. The plan would be to recover the lifting bodies from the ocean and break them up for the (high-value) pure titanium they would contain. There may be a time when the economics of that process will be favorable, but I would be reluctant to invest in a titanium import firm myself.

The "products" needed at L5, and available from Earth, will change as the communities develop from one bare outpost to a thriving, booming frontier settlement. In the beginning, L5 will need machines, tools, computers, and almost every other piece of complex equipment both for productivity and for life support. Carbon, nitrogen, and hydrogen from Earth will be needed until the time when the asteroids can be mined.

We should recall the fact that the velocity-intervals to L5 from Earth and from the asteroids are nearly equal. For that reason transport costs from Earth and from the asteroids may be comparable for a time. There may be then a period in which economic competition will tend to drive down freight rates for carbon, nitrogen, and hydrogen both from the asteroids and from Earth, although eventually transport from the asteroids should prove cheaper.

For a period of many decades during which the initial beachhead in space is expanding toward a mature community, L5 will need people, at a rate far more rapid than natural reproduction could supply. During all that time, the L5 communities will need to bring people from Earth, and we can expect to see, as we have in the case of Australia, a period in which free passage, initial personal "grubstake" capital, and perhaps initial free housing will be offered by the L5 communities as inducements to attract new immigrants from Earth.

The existence of those several components of a two-way trade, in which both sides would benefit, should help to

251

maintain a peaceful relationship between the L5 habitats and the nations of Earth. If irritations and misunderstandings do appear, as is inevitable in human relationships, fortunately neither side will be likely to risk a complete breakdown of trade; the price of serious conflict will almost surely be too high.

Though some items may be traded only for a short period, through much of the next century the need for additional energy here on Earth probably will ensure that L5 will have markets for new satellite generator capacity, and the communities' need for immigrants will probably continue about equally long.

Ultimately, if the L5 civilization nears maturity and the earth's population is stabilized, we can expect, in analogy with similar situations on Earth, that a two-way tourist trade will become an important part of the economic picture. We can be almost certain that such will be the case when we realize that in each passing decade the cost of transport, in constant dollars, will decrease as technology advances.

It has been said that new wealth requires three components: energy, materials, and intelligence. At L5 the source of materials will be inexhaustible at least for several centuries, and the source of energy will be reliable and effectively limitless for the next several billion years, to the best of our present knowledge. The third component is the human organization of machinery and human effort in a productive way. Productivity can be described by the ratio of output products to the input of human labor. If measured in nonmonetary terms (tons per person per year) the ratio automatically takes into account the effects of inflation.

For many centuries productivity was static, held down by the limitations of a hand-tool technology and the energy resources of human and animal labor. Then, with the industrial revolution, productivity began to increase. In the modern industrial societies of North America and Western Europe that increase has averaged between 2 and 3 percent per year. (It has been argued that in a pure capitalist economy, without government regulation, the interest rate on capital should be set at the same figure. Inflation, now several times higher than the productivity increase rate, adds to both productivity and interest rates in a way that tends to conceal the underlying real changes.)

Individual wealth is proportional to the productivity if government does not absorb a greater fraction of the total wealth as time goes on. A productivity growth rate of 2½ percent is enough to double real (uninflated) per capita income within less than thirty years. As we view the goods and services available and normally used in the western world today by people a generation younger than ourselves, we see that indeed our area has experienced at least a doubling of the real income in a time of just one generation. In space, although not on Earth, it is conceivable that such an increase in productivity could continue for a very long time. In the U.S. at present the annual family income is near $15,000. On Earth the limits on energy and materials are already putting the brakes on the increase of per capita income, but in space we can anticipate that by the year 2100, at a continuing growth rate of 2.5 percent per year, the average family income could reach the equivalent of more than $300,000 per year in noninflated 1975 U. S. dollars. Logically, that increase can only occur if available energy also increases, to a total of about two hundred kilowatts per person in a space society of the year 2100. Some of the amenities which we might consider for the end of the next century will not be energy- or materials-intensive. Perhaps the outstanding example of a sophisticated, energy-saving amenity is the electronic computer. Almost certainly by 2100 computers will reach a level of capability so high that nearly every common, predictable task will be computer-controlled, and will be carried out by machinery which in its turn will have been constructed in factories requiring very little human intervention. Other amenities will not be so economical of energy. Long-distance transport, for example, even in space will require a certain amount of energy. Logically, we can expect that by 2100 ordinary people living in space will take for granted the availability of inexpensive transportation, energy-intensive, which will give them tremendous freedom of movement over great distances at speeds of several thousand miles per hour. A two-dimensional array of space communities large enough to house the equivalent of Earth's present population, each person having at his service two hundred kilowatts of electrical energy derived from solar heat, would extend over less than 3,000 kilometers. Given enough energy, in space a normal cruise speed of 3,000 kilometers per hour would be quite practical, for an engineless vehicle accelerated by an electric motor. The equivalent of a whole world in diversity of population, climate, and landscape would therefore be available to a resident of space in the year 2100 within a travel time of an hour or less.

As the real income of the settlers in space increases, it seems unlikely that the residents of L5 will choose to remain in the rather cramped surroundings of the early habitats. On Earth, we are accustomed to the idea that with every passing year open space is enclosed, shopping centers spring up on once-green meadows, and the pressure on wilderness areas increases. At L5, where the rate of construction of new land areas will be limited only by productivity, we can expect that over the course of a century the population density can go down rather than up, whatever may be the absolute population size and its rate of increase. We can estimate roughly the population density of a new space habitat built in the year 2100 by taking present-day figures for the productivity increase rate and the world population growth rate. (Let's hope that's an overestimate; if it is the answer will be better than we're now calculating.) Taking the present U. S. value for the fraction of the population that's employed (around 40 percent) and assuming that a quarter of the workforce is employed in the construction of new habitats, we find that each new colonist of the year 2100 could have an "allotment" of almost two thousand tons of structure.

To see how far this would go, we need a model. "Island Two" will serve; each such Bernal Sphere would have a structural mass of several million tons. Putting the numbers together, each Island Two, with almost seven square kilometers of living area, would be occupied by only one

254

small village of twenty-six hundred people. Country living indeed! In space, all agriculture and industry would be located in additional area outside the living habitats, of course, so the L5 land area would be fully available for living space, recreation, and regions allowed deliberately to run wild (much of what we now call "wilderness recreation" area here on Earth was logged off or farmed less than a century ago, so the notion of a deliberate wilderness should not be strange to us). Even before correction for agricultural and industrial areas, the density would be comparable to those of some of the countries in Western Europe (the Netherlands have one thousand people per square mile, and Italy four hundred and fifty, with all its agricultural, industrial, and mountain areas included in the ratio).

On Earth, even with the assumed success of population-control programs, the total population will rise to at least ten billion some time in the next century. On the average we should assume then that population densities here will just about triple, until substantial emigration to the space communities takes place. Crowding, already severe in some areas on Earth, can be expected to get worse. In contrast, if we follow the population density projection for L5 another century into the future, we find for the space habitats a density less than a third of that now in mountainous, pastoral Switzerland, and considerably less than the average that the whole Earth will have by the early 1990s.

With increasing automation, it seems likely that the "standardized" portions of a new habitat—outer shell, mirrors, shielding, heat-radiators, and other externals—eventually will be constructed almost entirely by automatic machinery. Human intervention will be needed in just those areas where creativity and imagination will be called for: landscaping, architecture, and perhaps new artistic specialties like weather design and creative ecology. It may be that a group of settlers taking possession of a new land area in a habitat built by machinery will prefer to do the finishing operations themselves; to add the human touch by landscape, architecture, and choice of plant and animal species. Their first years could be spent in a way

255

Island Two habitat of the year 2100,
mainly forested, gives refuge for endangered species.

similar to those of our pioneer ancestors; each passing year
would bring a sense of accomplishment and the pride of
putting an individual stamp on home, garden and forest.

Specialists argue about the reasons for inflation; even
now, after many decades of effort and study, economists
are not in full agreement about its causes. The simplest of
all explanations is still in as good favor as any of the more
complicated: that inflation is caused by "more and more
demand chasing fewer products." A number of the factors
which may drive that supply/demand spiral on Earth
would not be present, or would be much reduced, in the
space environment. As noted before, energy costs at L5 can
be expected to decrease continually with the passage of
time, because the source will be free and limitless, and
technological advances can only serve to improve the

256

efficiency with which that solar energy is converted to usable forms. Once the asteroids become reachable for mining, every chemical element will be available in abundance, and the solar-driven transport systems for returning those elements to their points of use can only improve in efficiency and decrease in cost as technical development continues.

Here on Earth there is an inflationary pressure of the classical "more chasing less" variety, which we can observe in action every day. As the population density increases, land prices are driven up inexorably. Each time a new housing development is opened, prices are at a minimum; then, as the number of vacant lots decreases, prices go up until the last few go at prices almost of the sellers' choosing. If one wants to see inflated land prices, one need only look at desirable places where zoning laws keep the number of new building lots strictly limited, and where there are plenty of rich buyers searching for land; Switzerland is an outstanding example.

In the space communities population densities should decrease rather than increase. There should be no shortages of energy or materials. Perhaps then in the space environment there will exist the best conditions for a noninflating economy. If, over a period of many decades, severe inflation continues even in space, then our descendants must conclude that the main causes of inflation are not material but psychological. Even in that area the space communities may be at an advantage. We know that a primary psychological reason for inflation is fear; fear that some necessity or a product not essential but highly desired is going to "run out," so that an unreasonable price for it is justified; the "stockpiling" syndrome. Under the conditions of the space communities, after the first decades of learning and growth, it will be relatively difficult to create in the minds of the settlers the conviction that something material will soon be in short supply.

More uncertain than almost any other prediction about the future is any statement on the long-term effect of the space environment on the length of human life. Even so, one can make a plausible case for the statement that human life will be extended in space, though it will take some time before the prediction can be checked.

First, the fundamental conditions for the maintenance of life should be at least as favorable in space as the average in desirable areas of Earth, and far more favorable than in the regions in which most people now live. Poverty is a killer, and the wealth of space should permit most of the total human population to escape from poverty. Atmosphere, temperature, and sunshine in space can be optimized for good health. Given the shielding which could be obtained by the proper use of industrial slag, the radiation intensity in a space habitat should be no higher than it is on Earth. The risk of accidental death should be lower rather than higher in space. What though of the elderly? Here on Earth, with age and the infirmities of age, the body must spend more and more of its reserves of energy in simply fighting gravity. In the institutions to which many elderly people migrate, a great deal of the equipment one sees is devoted to the single task of assisting the body in its eternal battle with gravity.

In contrast, we can imagine that in a space habitat anyone with difficulty in walking will spend most of his time at a high elevation where gravity will be reduced; those who would be confined to bed on Earth could have freedom of movement in a region of near-zero-gravity.

Cardiovascular ailments are among the major causes of death for the elderly. In space we can expect that people with problems of circulation can move to low-gravity regions, and there enjoy freedom of movement and moderate, nontiring exercise. In summary, it seems quite possible that people in a space habitat will live to a greater age than they would on Earth. Perhaps it is even more important that their later years could be spent in conditions of far greater freedom and independence than their physical condition would permit them on our planet.

In the earliest of the technical notes on the modern development of the humanization of space I commented on the possibility of reducing the population of Earth by migration, perhaps by the middle years of the next century. [2] In doing so I emphasized, as I always must when the topic is raised, the difference between possibility and prophecy. If human migration into space does occur, it will certainly have the *power* eventually to permit such emigration, as you can prove with even the simplest of pocket calculators, using the numbers I've given in the last few pages for the mass of an Island Two habitat, and taking its population as 140,000. You need one more input: the fraction of the labor force engaged in habitat-construction; let's take that as half. We're being a bit conservative by assuming an Island Two geometry; Island One weighs less than half as much per person, as far as structure is concerned. Even without allowing for any productivity increase over the twenty-five tons per year that's now common in heavy industry on the Earth, you'll find a doubling time of only seven years for new land area in space.

I've assumed that the present-day value of productivity will still be appropriate to the year when the first Island One is finished. We'll take that as our "time zero." There might be a "dedicated" period of intensive construction after that time, as many of the nations of the world hasten to gain a foothold in space. In that pioneering era most of the space-dwellers (perhaps four-fifths) might be employed, and two-fifths of all their products might be new habitats rather than such things as satellite power stations. In that case, the doubling time for land area in space would be only two years, and in just eight years there might be 160,000 people living in space.

Let's trace what would happen if then the employed fraction dropped to the U.S. average, the colonists switched to building the larger Island Two habitats, productivity continued its present slow rise and—just as an exercise—all the output productivity in space went into new-habitat construction. How fast then could the population in space grow? (Notice I'm saying could, not will.) Again it's easy with even the smallest pocket calculator, and the answers are:

YEAR	POPULATION
10	290,000
15	1.5 million
20	9.2 million
25	68 million
30	631 million
35	7.3 billion

Before challenging these numbers, note that they're based on a continuation of the present slow rate of productivity increase; without that, the time scale would be somewhat longer, though not much: the population shown in the table for year 30, for example, would be reached about five years later.

The point of this calculation is that the productivity that we have achieved already on Earth, when employed in the energy-rich, materials-rich environment of space, could lead within less than two generations to a production-rate of new land area great enough even to accommodate the population increase rate of Earth. If the number of people on our planet rises to ten billion, and if its rate of increase goes unchecked, that rate of increase will be 200 million people per year. In the table, it would require only thirty years from the completion-date of the first community before new lands would be increasing more than fast enough even to cope with such demands.

That exercise is not presented as an "optimum scenario." Indeed, I would much prefer to see our growth rate here on Earth decrease with time; but I would like to see it decrease for the right reasons: security, a decent standard of living, and free choice; not for what seem to me the wrong reasons: legal or economic coercion.

The second part of what one might call the "emigration problem" is transport: is it reasonable to consider a transportation system with the capacity to cope with such rates? Again surprisingly, the answer seems to be yes. In Chapter 10 I described a relatively near-term vehicle system based only on the technology we believe we now understand. The fleet of vehicles I described would be capable of carrying about five hundred thousand people in one year from the earth to L5. In the "fastest possible buildup," that emigration rate would be reached in about year 15 from the beginning of the "Island Two" era and a rate of two hundred million per year would be reached about fifteen years later.

To accommodate that higher rate, we'd like to have power supplies for shipboard use with a mass just under a ton per megawatt. That could come about either by several decades of development of solar-cell technology, or by the use of microwave or laser transmission of power in space. With such performance the round-trip travel time for a large ship powered by a mass-driver engine could be as little as twelve days, with the outbound trip taking only three and a half days—less time than is required by the fastest ocean liner for the Atlantic crossing. If each ship were to carry 6,000 passengers, a modest increase in capacity over a fifteen-year period from the time of the *Tsiolkowsky* and the *Goddard*, then about eleven hundred ships in all would be needed. That's comparable to the number of large ocean vessels that now sail the waters of Earth. If we check the productivity required for the con-

261

struction of eleven hundred large spacecraft we find that their total mass would be some ten million "deadweight" tons, and that they could be built in three years by a work force of fewer than 0.1 percent of the population that L5 would have in year 25.

Transport from Earth to low earth-orbit, during the same era, would presumably occur in vehicles with passenger cabins as large as those of a Boeing 747. Compared with the capabilities of the present space-shuttle, that's an increase over a time of around fifty years that's much more modest than our own experience in aircraft: from the 24-passenger DC-3 to the 400-passenger 747 in only thirty years. The trip from Earth to low-orbit would take less than half an hour, whatever the vehicle size, and for a round-trip time of four hours the transport demand could be met by a fleet of less than two hundred vehicles. That's only a tiny fraction of the number of large aircraft (about four thousand) already in the world's commercial jet fleet.

Ticket costs calculated by the same methods used earlier would be about $4,500 per person in today's dollars; comparable to the present cost of a round-the-world trip, and equivalent to only a few months' earnings under the conditions prevailing in the communities.

From the industrial societies of North America, we pour into the atmosphere each year about ten tons of combustion products for each member of the population. Over a lifetime each person is therefore accountable for more than six hundred tons of combustion gases and smoke. By contrast, the fuel used to lift an emigrant to low orbit from the surface of Earth, by vehicles no more advanced than those of the present day, would be less than three tons—only as much as he would be using in a four-month period on Earth. Once an emigrant left, the corresponding burden of his energy usage on Earth's resources and atmosphere would be lifted, permanently except for his later visits to the home planet. If the traffic to and from space ever reaches the frequency given in the example it will be very important to design engines for clean-burning fuels, and to pay special attention to the delicacy of the atmosphere's ozone layer. There will be at least forty years of time to study the problem before it will be necessary to solve it, so

I think we can conclude that there are no serious obstacles to handling even as great a traffic volume as has just been calculated.

When we consider this possibility of reducing the population of Earth by emigration, it is important to distinguish possibility from prophecy. As we have seen, the combination of technique with natural growth in capability would have the *power* to permit such emigration. Whether or not large-scale emigration will occur will depend on how badly it is needed, and on how attractive the space communities become. With four billion people, Earth is already overcrowded in many areas; many would choose to flee Earth if it had ten to fifteen billion.

The availability in the space habitats of high-paying jobs, of good living conditions, and of better opportunities for children may stimulate the emigration of a considerable segment of Earth's population even if overcrowding on Earth is less serious than now appears likely. In the long run, because of the availability in space of unlimited cheap energy, of abundant materials, and of efficient combinations of attractive living area with nearby industry, I suspect that Earth-based industry will be unable to compete economically with space-based industry. If so then, as has occurred many times within Earth's history, people will follow the availability of jobs, and that will mean emigration.

A nonindustrial Earth with a population of perhaps one billion people could be far more beautiful than it is now. Tourism from space could be a major industry, and would serve as a strong incentive to enlarge existing parks, create new ones, and restore historical sights. The tourists, coming from a nearly pollution-free environment, would be rather intolerant of Earth's dirt and noise, and that too would encourage cleaning up the remaining sources of pollutants here. Similar forces have had a strong beneficial effect on tourist centers in Europe and the United States during the past twenty years. The vision of an industry-free, pastoral Earth, with many of its spectacular scenic areas reverting to wilderness, with bird and animal populations increasing in number, and with a relatively small, affluent human population, is far more attractive to me than the alternative of a rigidly controlled world whose people tread precariously the narrow path of a steady-state society. If the humanization of space occurs, the vision could be made real.

Science-fiction writers are fond of assuming such conveniences as faster-than-light travel ("Warp Factor Six, Mr. Sulu"), suspended animation, and teleportation. When speculation is involved I find it more challenging to see just how far we could go *without* assuming any science beyond that of our own time.

Earlier I described an asteroid-voyaging research vessel capable of roaming the inner solar system with a laboratory-village of several hundred people. In space the limits on the size of a vehicle would be far more relaxed than they are on Earth, and there is no reason in principle for not considering much larger mobile objects. A habitat the size of Island One could be equipped with a solar-electric propulsion system of the kind described in Chapter 11. Human populations of 10,000 have existed in isolation for periods of many generations, within the history of our planet; that number is quite large enough to include men and women with a wide variety of skills. Space dwellers will be well equipped psychologically for distant voyages, and a few decades after the beginnings of the human settlement of space there may well be large groups of people roaming the outer reaches of our solar system, on long-term missions with a scientific purpose. Such groups could be connected intimately to the rest of human society by television and radio, so there would be no reason for them to remain isolated unless they chose isolation for reasons of their own. Even at the distance of the planet Pluto, the most distant known member of our family of planets, new examples of the visual and musical arts could be received with a time-delay of only a few hours.

We can estimate the approximate limits to which a roving space habitat could go by assuming that its inhabitants would want full Earth-normal sunlight, that the total land area for habitation and for agriculture would be that of "Island One" that a solar-electric generator supplying

264

the habitat with the present U.S. total per capita energy usage would be a desirable accessory, and that the mass of a collecting mirror to concentrate sunlight should not be more than double the mass of the habitat. The corresponding limit of distance, if the mirror averages several wavelengths of light in thickness, is roughly four light-days: about ten times as far out as the orbit of Pluto. That limit, approximate rather than exact, corresponds to a kind of "continental shelf" for our solar system; beyond it lies the abyss of interstellar space. Within that limit, though, it seems that there is no reason why a roving community should have to endure conditions any less luxurious than those of the habitats nearer Earth.

Research town of Island One size, collecting faint sunlight, operates at the edge of the "continental shelf" beyond Pluto.

A spacefaring laboratory would be a gossamerlike affair, with a huge paraboloidal mirror. At the center, spiderlike, the shielded habitat itself would rest, absorbing the solar energy collected from several thousand square kilometers of area. Within, I suspect that the traveling laboratory would be landscaped in a manner expressive of the inhabitants' need for the psychological solace of lush vegetation. Long-time inhabitants of such communities would probably develop a passion for gardening, not only for flowers but for unusual vegetables and spice plants.

Of the inhabitants, in that case necessarily limited to a constant population, about a quarter would be of school or college age; enough to require a small university. Half the population would be within the normal working years, and of those people half again might be needed for all the services of the community: teaching, agriculture, maintenance, engineering, navigation; people to operate stores, printshops, cinemas, hospitals, libraries, and restaurants. The replacement of durable goods by up-to-date equipment manufactured according to plans radioed from L5 might occupy another fifth of the work force. The remainder, perhaps 2,000 people, could be directly engaged in research: planetary astronomy, geology, geophysics, interstellar astronomy, and the operation of long-baseline radio telescopes in partnership with laboratories near Earth. A laboratory of that size would be comparable to the whole faculty and staff of a medium-size university or a large national laboratory of the present day; it would be quite large enough to carry out, over a period of years, thorough and systematic explorations of the outer planets, sending down manned and unmanned probes to the planetary surfaces for short excursions.

Living in a community like that would be rather like living in a specialized university town, and we could expect a similar proliferation of drama clubs, orchestras, lecture series, team sports, flying clubs—and half-finished books.

Guessing at the deep-space activity within the limits of our solar system's "continental shelf" during the next century, I suspect it will be confined to asteroid-mining, to communities roaming the solar system for research, and to small fixed research colonies on habitable planets. The bulk of the human activity, in my guess, would be concentrated in the region near Earth and in the asteroid belt, and would be linked by a communications network with delay-times given by the speed of light, and therefore no longer than about half an hour.

Our first good look at nearby star systems probably will be through large, composite (many-mirrored) telescopes based at, but not on, space communities. Perhaps our descendants will find, some time in the next hundred years, that a star within a few light-years of our own is sufficiently interesting to warrant a closer look. That might occur if the star were proved by telescopic observation to have planets, for example. In that case a robot probe could be dispatched, on what would be a voyage of many years. The most economical way to gain information about another star system would be through a "fly-by"; the probe would use all its store of energy and all its reaction mass simply in accelerating, in order to minimize the travel time. It would plunge through the target star system at a speed of perhaps a tenth that of light, and in a few hours would gather all the information that its sensing elements could retrieve. Then, over a time which might well be measured in still further years, it would radio back to its human masters all the information collected during its brief hours of intense activity. As we view the rapid development of computers and miniaturized electronics, it seems safe to state that a century from now a robot probe could be far more reliable and sophisticated than any possible human crew, so our first close look at another star system is most unlikely to be through human eyes.

Could a space community some day ever venture beyond our continental shelf and embark on a trip to another star? If the community were large enough to constitute a complete society, and if the social stability of an isolated large group turned out to be great enough, such a voyage certainly would not exceed the bounds of physical possibility. But for that we must carry our speculations well

Unlike all else in
book, this drawing
goes beyond technology
of year 2000, depicts
interstellar vessel
(twenty-first and
-second centuries).

beyond the limits of present-day technology. Vessels limited to engines that could be built relatively soon, within the next few years, and which would use solar energy to maintain earthlike conditions, would be limited to distances of a few light-days from the Sun. For interstellar distances a source of energy would have to be carried on board. Though the television series *Star Trek* assumes much technology contrary to physics as we now understand it, some of its technical paraphernalia make sense within the limits of our present knowledge. In particular, the "matter-antimatter pods" about which Engineer Scott is always so worried are quite reasonable, assuming the technology of a century or two from now. Particularly in space, without gravity to bother us, it would be possible to build up a quantity of antimatter. The cost in energy would be enormous, and at present our methods for producing antimatter are primitive and inefficient, but there is no reason why they should always remain so. The most convenient form in which to carry antimatter would be liquid or solid; frozen antihydrogen, at a temperature of only a few degrees above absolute zero, would be a good candidate. Its atoms would consist of antiprotons around which positrons would circle.

Returning to the example of the "Island One" community, equipped for distant voyaging, we can imagine its mass-budget for a mirror being replaced by an equal quantity of frozen hydrogen and antihydrogen. The antimatter could be maintained, in the absence of gravity, by electrostatic fields requiring no direct physical contact. It could be protected from the ordinary-matter cosmic radiation and from dust particles by a thick shield of ordinary matter, and in principle should last a very long time. When we calculate how long a space community could exist, in Earth-normal energy conditions, on stored antimatter fuel, we find that it could last for several billion years! Certainly plenty of time for an interstellar voyage.

In the second chapter of this book I had occasion to quote the conclusion of Professor Richard Heilbroner regarding the outlook for humanity if its only ecological range remains on our planet. "If then, by the question 'Is

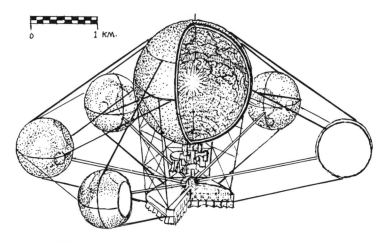

Interstellar vessel of Island One size,
ion-driven. Central arc substitutes for sunlight.
Reaction of matter and antimatter gives
power (twenty-first and -second centuries).

there hope for man?' we ask whether it is possible to meet the challenges of the future without the payment of a fearful price, the answer must be: No, there is no such hope."

Those of us who have enjoyed since birth an adequacy or even an excess of material comforts are the first to describe them as of secondary importance; not so the large fraction of the world's population which moves in pain and poverty from birth to the grave. As we view the problems which now face humanity as a whole, it seems least excusable that now, late in the second century of the industrial revolution, so much of humanity remains in need of even the essentials for health and a decent life. Clearly, given a worldwide dictatorship of unchallengeable military strength and egalitarian outlook, the wide disparities in the wealth of nations and individuals could be much reduced. By my standards, to accept such rule, even if there were any realistic way in which it could be brought about, would be indeed to pay "a fearful price."

As we look back on the times of which the human race is most proud, it is hard to escape the conclusion that they

271

were times of diversity, competition, unpredictability, and considerable confusion. We still recount in admiration and pride the philosophical, literary, and dramatic accomplishments of the Greek city-states; in that period were born many of our most cherished concepts of freedom and individual worth. Is it pure accident that the classical era was also a time of great diversity and of disparate, often conflicting ideas from small communities of no more than a few thousand individuals?

I wonder, too, at the age of darkness which followed the monolith that was Rome. There, if ever in the distant past, an organized state arose with the power very nearly to take over the entire world; with ideals and a concept of civilization not completely abhorrent even to our modern view. Yet that brief period of supranational organization was followed by many centuries during which, as we now see it, little of what we think of as "progress" occurred. Is there something in the concept of universal organization that is basically alien to humanity? Something against which the human spirit rebels? Perhaps so; the next period in which we as humans take great pride is the Renaissance and the age of exploration; certainly a time of great differences, great uncertainties, and unprecedented mobility.

As we view the next decades with the new option of the humanization of space, of one thing we can be sure: those years will be unpredictable. New possibilities have appeared, and with them literally a new dimension in which humanity can move. From Flatlanders, we have suddenly the opportunity to become the inhabitants of a three-dimensional solar system. Clearly our first task is to use the material wealth of space to solve the urgent problems we now face on earth: to bring the poverty-stricken segments of the world up to a decent living standard, without recourse to war or punitive action against those already in material comfort; to provide for a maturing civilization the basic energy vital to its survival.

These are the immediate problems, and I have attempted to show how these problems can be solved by ourselves; they do not require supermen with a more than human capacity for organization, cooperation, and self-denial.

It may be argued that the exploration and the settlement of space is no more than a "technological fix" for problems that should be solved on a higher, more intellectual plane. Yet by our evolution we are closely tied to the material world; we are the descendants of the survivors, from many generations during which the maintenance of life was a struggle every day with the material world. Our history does not suggest that we are well-suited to changing, overnight, to a species disinterested in material well-being, with paramount concern for humanity as a whole rather than for a narrower group. Indeed, our loyalties are first to those few individuals to whom we are linked by close ties of genetic relationship; only with an effort do we extend our concern to the town, the state, the nation, and the world. As a species, we have solved our problems by technical means for millennia, and it would be surprising indeed if we could change our character so completely as to abandon the methods by which we have survived.

Earlier I contrasted the new ideas in this book with the philosophies of the classical Utopians. Will the space communities be free of conflict, free of misery, free of sadness? Certainly not, as long as they are human. Rather, we should hope that they will give added opportunity for that most elusive of human occupations, so fundamental as to be written into our Declaration of Independence: "The pursuit of happiness." Our country has not survived its first two centuries on the basis of promised happiness; rather, on the promise that the search for happiness could go on. I hope, and I think, that those people who have taken the concept of the humanization of space so much to their hearts have not done so in a misguided expectation of perfection. If their letters and their conversations are any guide, they appreciate how difficult will be the conditions and the challenges that they will meet in a new age of exploration and discovery. Yet even the opportunity to try new ideas and to break out in new directions is more than could be hoped for in a world forever limited to the confines of our planet. It is, after all, only a few thousand years, perhaps a hundred human lifetimes, since human-kind first abandoned the nomadic existence of the hunter for the stability of farm and cottage. No wonder, then, that

273

there should be, deeply rooted even after those years, a need within us to feel that boundaries can be broken and new paths explored.

What of the arts, and of letters, in a new period of expansion of the human spirit? Creativity is the most difficult of the human attributes to predict, but it is at least hopeful that the age of Columbus and of Drake was also that of Michelangelo and of Shakespeare.[3] On a more homely level, occupations that have a flavor of openness and nomadic existence have always been celebrated in our romances; in modern popular song the ever-moving truck driver has taken the place that was occupied a century ago by the cowboy. In the challenge of the first outpost in space, and of the voyages that may be undertaken by those who travel with their families to the lonely asteroids, there should be matter enough for song and story.

As we consider the human prospect in space, we know that where people are involved there will always be the potential both for good and for evil; yet there seems good reason to believe that opening the door into space can improve the human condition on Earth. Relieved even a little from the drive to struggle with other nations for the diminishing resources of our planet, we may hope for a more peaceful future than will otherwise be our lot. Generosity toward the Third World, in its attempt to avert famine and to take its place among the community of nations, seems more likely to be shown if that generosity can derive from new, unlimited resources rather than from those we already find to be in short supply.

More important than material issues, I think there is reason to hope that the opening of a new, high frontier will challenge the best that is in us, that the new lands waiting to be built in space will give us new freedom to search for better governments, social systems, and ways of life, and that our children may thereby find a world richer in opportunity by our efforts during the decades ahead.

APPENDIX 1

TAKING IT TO THE PEOPLE

In the late 1960s disenchantment with the sciences had become general, and massive budget cuts in research had already taken place. Yet in that same period the Apollo Project, begun several years earlier at a time of confidence in American power and capability, reached its fruition with the first human landings on the Moon. As others have said, our age may be remembered for no other accomplishment than that first great climb from Earth's surface to another planetary body.

In that same period the horrors of the war in Southeast Asia had provoked a revulsion against authority and against technology on American university campuses. At many colleges there were active riots, and in extreme cases acts of violence resulting in death. Princeton remained relatively calm, but even in our quiet backwater there were meetings and demonstrations against academic authority. Students who sensed that they had talents in science or engineering were on the defensive, accused by their colleagues of being "irrelevant," or in another catchphrase of the time, "counterproductive."

At the height of that period of campus unrest it became my turn to teach the largest of our first-year physics courses. The level of the course was relatively stiff, requiring calculus, so its clientele included prospective physics and math majors, engineers, a sprinkling of other would-be scientists, and an occasional brave premed, willing to risk lower grades to learn physics at a level higher than required by the syllabus.

With the prospect of a year spent in teaching double the normal amount, it was natural to choose the reorganization and modernization of the course as a challenge both diverting and, hopefully, useful. Some of the changes were minor: abandoning the traditional blackboard in favor of an overhead projector that would permit standing close to the front row of students, facing toward rather than away from them. We did away with homework handed in for grading weekly, in favor of "learning guides," programmed instruction booklets in which each student could find help and guidance even when studying on his own. For rapid feedback of information to the students, we returned to the old-fashioned custom of weekly short quizzes, promptly graded.

At Princeton, as in most research-oriented institutions, there had been wide variation in the help available to students outside of class hours. Some members of the

276

teaching staff, whose research depended on machines thousands of miles from the university, were often unavailable even though well-intentioned. To alleviate that problem, we began a system of cooperative office hours, so that each hour of the day during which a student might reasonably seek help would be "covered" by a staff instructor assigned to that particular time.

To unify the course a theme was necessary for the year: it was easy to choose: *Apollo 11*, the first lunar landing mission, had flown successfully only two months before, and *Apollo 12* was due to lift off only two months after classes began. The Apollo Project, though already under heavy fire for its "irrelevance" to the needs of the inner cities, was exciting and offered many possibilities for illustration in a freshman physics course. Following that plan throughout the academic year 1969–70, each area of physics in the course was illustrated with examples from the first series of human voyages into space: force, energy, and momentum: celestial mechanics, thermodynamics, electrical theory. For one of our "laboratory" sessions a simulator was set up, based on an early, primitive version of an electronic desk calculator, and with it the students participating were able to carry through the simulation of a landing on the moon. When their understanding of the optimum directions and timings of rocket firing were faulty, they found themselves running out of fuel while still three hundred feet up, and tension ran high in the laboratory when that happened.

In any large course, one must aim the instruction at the middle of the class, and then make special arrangements for those who are either much slower or much faster than the average. In the case of Physics 103, after the changes just described the slower students were being given the help they needed; it remained to make special effort for those whose preparation, natural talent or motivation was so much above the average that they were being challenged insufficiently by the regular teaching. For the first months of the course, before the work overload piled up too high for all the students, I held a small, informal seminar, in an attempt to alleviate that problem.

Given the peculiar problems of 1969 on a university campus, it seemed that we should attack the question of the place of the scientist and the engineer in the society of the next decades. Clearly, the days of blind trust in science and in progress were past. Not only because of the real needs of the world outside, but because of the self-doubts and questionings of the would-be scientists themselves, it was important to examine problems relevant to the issues of the environment, of the amelioration of the human condition, and of the interaction between science and society.

The traditional view of the scientist, and the value-system which has been associated traditionally with excellence in the sciences, is oriented to specialization. The trite phrase "knowing more and more about less and less" sums up much of this view, and until very recently scientists who crossed boundaries between fields of specialization were viewed with considerable suspicion by their colleagues. One cannot dismiss that attitude lightly. It is easy to make mistakes when working in an unfamiliar field, and far too easy to work in several fields while doing a really first-class job in none. There are unhappy case histories of good scientists brave enough to cross the boundaries between scientific disciplines, who have discovered, usually only painfully and after years of wasted effort, that they could not gain an adequate degree of mastery of the new subject in the time available to them.

Yet the students of 1969 were seeking for relevance in their own careers, groping for ways in which their natural talent for technical subjects could benefit their fellow human beings. Above all, they sought to avoid a narrow specialization which would place them in that sad category described by Dickens:[1]

". . . The misery with them all was, clearly, that they sought to interfere, for good, in human matters, and had lost the power forever."

In our seminar, held each week and attended usually by some eight or ten students, I hoped to discuss large-scale engineering problems that would combine several attributes: they would have to be broad enough in scope to be challenging, and their solutions must benefit a broad spectrum of humanity, especially those disadvantaged by accident of place or situation of birth. If those problems were to be attacked by the students of our seminar, their solutions must not require materials, techniques, or engineering knowledge beyond the level of the 1970s or early 1980s. As it turned out, once a first problem was chosen for discussion, it so occupied us that we never had time for a second.

Often people have asked why I picked as our first question: "Is a planetary surface the right place for an expanding technological civilization?" There is no clear answer, except to say that my own interest in space as a field for human activity went back to my own childhood, and that I have always felt strongly a personal desire to be free of boundaries and regimentation. The steady-state society, ridden with rules and laws, proposed by the early workers on the limits of growth was, to me, abhorrent.

The level at which we could attack this question was necessarily modest. The students of Physics 103, in October of their freshmen year, were only four months past graduation from high school.

First there was the issue of energy: in space, solar energy would be available full time. We could imagine no cleaner, more inexhaustible, or ultimately cheaper source of energy for a society which by our assumption would be expanding and growing in technological capability, if not necessarily in population. The possibility of the colonization of planets other than Earth could be dismissed quickly, for another reason besides the unsuitability of a planet to solar energy use. The land area available was insufficient; the use of the Moon and of Mars would hardly more than double the land area available to us as a race, and at our present rate of expansion that increase would be used up in a mere thirty-five years.

What about colonies in free space itself? First there was the question of their possible size. From the beginning we were thinking in terms of something earthlike, not just a space station. There must be the possibility of ordinary human life, complete with gravity, atmosphere, sunshine, growing plants, trees, and animals.

It was clear that there could be only three basic geome-

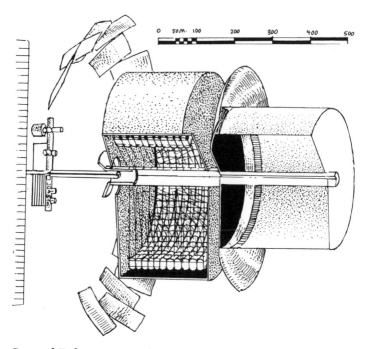

Crystal Palace or "Hatbox" gives
maximum land area at normal gravity, for
minimum mass of structure and shield.

tries for a vessel in space, containing an atmosphere and
rotating to provide an artificial gravity: the sphere, the
cylinder, and the wheel (toroid). The last had been discussed thoroughly during the 1950s, and seemed to us
more appropriate for a space station than for a mini-
world. The sphere seemed to us less desirable than a cylinder, because we sought to maximize usable land area at
a gravity near earth-normal.

Our first assumptions seem naïve in retrospect: we
thought in terms of full normal atmospheric pressure, and
of a soil thickness of about five feet, far more than is used
by most growing plants. Even so, our first calculations
showed us that a steel shell rotating to provide earth-level
gravity, and loaded by that soil depth and atmosphere,
could be built in a size as large as several miles in

diameter. That first numerical result surprised us, and stimulated our asking further questions.

What about room for expansion? At that time we had only a hazy idea of the total volume of material available in the asteroid belt, but it seemed that asteroidal material would suffice to build space colonies with a total land area at least many thousands of times larger than that of the earth. It was Freeman Dyson, more than a year later, who pointed me toward that mine of information, Allen's *Astrophysical Quantities*, where the correct numbers could be found.[2]

It remained to find a way in which sunlight could be introduced into the rotating cylinder, preferably maintaining the visual effect of the normal disc of the sun in the sky, and its slow passage across the sky during the course of each day. Perhaps it was the fourth or fifth of our seminars to which I brought a model, made out of bits of paper, tape, and plastic, of a cylinder subdivided into six segments, three of them transparent, into which sunlight could be brought by external planar mirrors.

My calculations, after the seminar ended, were carried out in occasional spare hours on weekends or late at night; often they were done at times when my schedule made it necessary for me to spend a day or so in another country, away from my regular research or teaching. The more the problems of setting up communities in space were examined, the more it seemed that reasonable solutions existed for each problem. That experience happens rarely to a scientist; in most cases a new idea is shot down quickly by the first few calculations. One learns to recognize the exceptions. There was a definite sensation of "deja vu." Thirteen years earlier, in 1956, it had been my good fortune to experience the same excitement, the same feeling of exploring a new logical path, when at that time I had begun to work on the possibility of storage-rings.

In 1956 I was twenty-nine, and had been at Princeton for two years as an instructor. At the invitation of Professor M. G. White I had chosen to work on the design of a large new accelerator for protons. The design process was fun; perhaps only in those far-off days of the mid-fifties, when physics enjoyed a high level of support and its practitioners were relatively few, could it have happened that someone so young could take a significant part in a large-scale problem of systems design.

In the midwest, Professor Donald Kerst and a large group had begun working on the theoretical possibility of building a special type of accelerator in which two beams of particles could circulate simultaneously, in opposite directions. Occasionally, in such a machine, collisions between those particles would take place, and those collisions would be the most energetic that could be achieved by man in the laboratory. The energy scale of those collisions would be so far above that which characterizes nuclear transitions that there could be no expectation (neither hope nor fear) that they would produce nuclear energy, as in a reactor or an atomic bomb. They would be used in pure research, to teach us much about the constituents of the neutron and the proton.

Unfortunately, the special machine demanded by the design work of Professor Kerst's group would have been massive and expensive, and would have provided only marginal access to the interaction site for experimental detection apparatus. As it stood, it appeared that the colliding-beam concept could be realized theoretically, but only at such expense and with such difficulty that it might never become usable in a practical way.

On looking at the problem it seemed reasonable to ask, "Is it necessary for the collisions to take place in the same machine in which the protons are accelerated?" Calculations at Princeton indicated that the two problems of acceleration and of storage could be separated. So began the modern development of what are now called "storage-rings." Some of the same ideas, in a form apparently not convincing enough to justify further investment of time, had occurred to a European engineer, Rolf Wideröe, during World War II. Wideröe's work, to which he called my attention several months after my publication of the concept, was buried in the form of a wartime German patent and, as far as I know was not subsequently published. William Brobeck, at the Berkeley cyclotron laboratory, reinvented storage-rings at about the same time that I did.

From the first concept to practical realization, in the

form of a high-energy experiment, took almost ten years of effort. For the construction project co-workers from Princeton and Stanford joined me and the first high-energy colliding-beam experiment was carried out by our group in 1965. It proved that the charge of the electron was confined within a tiny volume, less than a thousandth that of the proton.

Even then it would have been impossible to imagine that the storage-ring concept, once so controversial, would pass in ten more years beyond the point of acceptance to the stage of near-universality. As of 1976 nearly all of the effort on new particle-accelerator design, in every country active in that field, is for colliding-beam storage-ring machines. Perhaps it was the experience of that previous transition from incredulity toward acceptance that encouraged me, back in 1969 and in the early 1970s, to continue working on space communities, another "crazy idea" which carried the same sort of logic. In 1969 as in 1956, "the numbers came out right."

Double-teaching and high-energy research were more than a full-time job in 1969–70, but as the numbers concerning space communities continued to work out well I became interested in communicating the work to others. At first, that communication was informal and casual: to my three children, in long walks through the woods near Princeton on brisk days in late autumn and early spring. It seemed important to discuss with my children a new option which might expand greatly the range of choice available to them during their lifetimes. Sometimes I would discuss my work with friends, but I was far too shy about working on such frivolous, easy, elementary physics to talk about the work with my colleagues at that time.

One evening at the home of a friend someone suggested writing up these ideas for a well-known literary monthly. There followed a pleasant interchange with the editor. He was interested and asked many questions, answered in a second draft. Then came a friendly letter of rejection: "I'm sorry, I'm fascinated by this, but I've asked ten questions and you've answered them, and now I feel like asking a hundred more, and I'm afraid that the process isn't going to converge."

In 1971–72 I continued to attempt to put the new ideas before a wider group for discussion. In doing so I ran into a phenomenon, well known to most aspiring writers, from which I had been spared up to that time: the turnaround time for rejection slips. In accord with the rules the article was submitted to only one magazine at a time.

There followed, usually, from four to six months during which the manuscript was tied up and could not be submitted elsewhere. I did not want to write my work in the form of science fiction, because I felt that such a "cop out" would put the concept in the milieu of fantasy, from which a transition to serious discussion would be far more difficult. It seemed reasonable to try science-oriented magazines which had published my work in the past. There were two such magazines, widely circulated, in each of which I had published two articles during the 1960s. The first and somewhat less scholarly of the two rejected my cautious letter of inquiry brusquely; the editor did not even want to see the manuscript. That at least was rapid turnaround. The editor of the second magazine offered to look at the manuscript and have it reviewed. By that time it was mid-1972, nearly three years from the time of formation of the basic ideas.

The second science-oriented magazine also rejected the manuscript, having been advised to do so by both of its reviewers. The editor was kind enough to send excerpts from the reviewers' comments. Their reasoning is of interest to recall. One reviewer went into a state approaching shock on learning of the new ideas: his argument could be summarized as "No one else is thinking in these terms, therefore the ideas must be wrong."

The second reviewer, more thoughtful in his response, made a number of assumptions which were reasonable enough within the mainstream of contemporary thought, but which happened not to apply to the new dimensions of opportunity opened by the space-community concept. By a curious chance, I was to meet and talk with that second reviewer several months later.

In thinking back to that period several years ago when it was so difficult to gain a hearing, I cannot help but go through a little of the introspection which is so common to men in their forties. At that age one is at the midpoint of a formal, salaried career, though some of us cherish the hope still to be productive in terms of writing and the interchange of ideas well past the time of retirement. One cannot help, at that age, looking back and attempting to find a pattern in one's own weaknesses and, one hopes, talents. In my own case it is easy to count weaknesses: there are so many of them. The talents are harder to find. As I compare my work with that of colleagues, I can find in my own a modest degree of the usual professional competence in the standard areas of mathematics, physics, engineering design, and so on, but can hardly claim more than professionalism in those areas. As I look back on two unique periods in which I initiated something that turned out to be worthwhile but which no one else pursued, I find that there were close similarities in the two developments. Of both storage-rings and the humanization of space it could be said that no great mathematical talent, no great height of theoretical abstraction, was involved. Rather, if any talent was involved it seemed to be that of finding easy ways of solving large-scale problems of systems design. In each case, the new way required going off in a new direction from the path followed by others. Both in the case of storage-rings and of space communities there were particular sub-system devices, critical to success, which I had to invent before the new synthesis could take place. In the first case, it was a device called a "delay-line inflector," which could switch a particle beam from one track to another within a small fraction of a millionth of a second, without spoiling the quality of the beam. In the case of the human settlement of space, the essential device seems to be the mass-driver, the machine for launching material from the surface of the Moon. Probably in both cases, and quite certainly in the first case, the "essential device" was only the first and easiest solution of a problem which eventually could be solved in several different ways.

This talent, if such it is, is not profound, but perhaps I can take comfort from the fact that it seems to lead toward developments which have some value.

In the summer of 1972 the problem of getting the new possibilities discussed was becoming serious enough to trouble me considerably. At that time, the rejection-slip process had stretched over more than two years, and the list of publications, yet untried, which would be suitable for a first paper on space communities was becoming rather short. My children and I spent a month together, camping in upper New York and New England, and in that month I experienced for the first time the wonderful feeling of release that comes from learning to soar, to fly a sailplane, with its three dimensions of freedom and its intimate interaction between the machine, the pilot, and the invisible, ever-active atmosphere outside. Returning from our flying and camping vacation at Elmira and Franconia, we stopped for a day to visit old friends, Brian and Joyce O'Leary. Brian and I had met five years before, in San Antonio, as finalists in a scientist-astronaut test series. Later, Brian had entered the astronaut corps but had subsequently resigned from it, and in 1972 was teaching astronomy at Hampshire College. Brian and another friend from San Antonio days, Professor George Pimentel of Berkeley, encouraged me to bypass the traditional process of academic publication. "Take it to the people," they advised, and Brian suggested that he arrange for me to give a lecture for students at Hampshire College that autumn.

When I returned to Princeton and the new semester began, another old friend, Professor John Tukey, gave me the same advice. We spent a long lunch at the faculty club, John selecting from his pocket phonebook names of people from many fields of academic expertise to whom I might turn for commentary on the new ideas. John is legendary at Princeton. In the days before the invention of electronic computers, the story goes that the academic class schedule at Princeton was made up each year by a simple process: John Tukey would lie down on a couch each morning for three days and someone would read to him all the conflicting schedule requirements of every course taught at the university. John would rest quietly, taking it all in, and then on the fourth day, never even using a pencil and paper, would dictate a complete class schedule for the following year, with never a conflict. These days for that

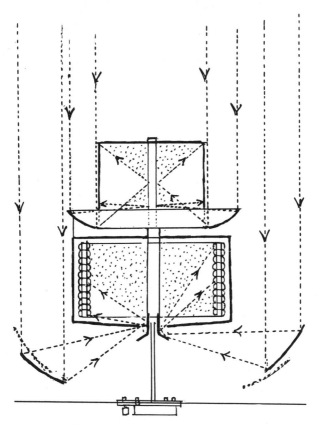

Sunlight paths for Crystal Palace.

task John has been "replaced by a computer" and thus has a few extra days each year to devote to his responsibilities as head of the Department of Statistics.[3]

Leaning back in his chair at the end of our discussion, and quietly staring at the ceiling as is his habit, John concluded with a remark that very kindly assumed for me a place beyond my aspirations: "Remember Goddard," he said, "and don't get discouraged."

In October of 1972 I was teaching double the normal amount again, in preparation for a semester to be spent wholly on research in high-energy physics, at Stanford University. The class schedule was rather tight, and I could lecture at Hampshire, without missing a class at

Princeton, only by spending all of one afternoon on the four-and-a-half-hour drive to Amherst, and on the following morning getting up at three o'clock to begin the drive back. Fortunately, schedules of that kind are common in high-energy experimental work, so I was well inured to them.

Brian had told his students of the lecture, and word-of-mouth had spread the information rather widely in the College; the lecture, starting at eight in the evening, was well-attended by students and faculty. Slides, made from my own crude sketches, were just enough to show people the essential ideas, and I spoke for less than an hour.

The response from the students was strong and positive, enough to give me courage to continue. Questions went on for an hour, and then one of my hosts, Dean Everett Hafner of the School of Science at Hampshire, got up and spoke to the audience:

"I just want to say," he began, "that when I first heard these ideas I thought they were crazy. Now on hearing this discussion I've changed my mind, and I'd like you all to know that.

"You should also know that the speaker must make a long drive tomorrow morning in order to be back at Princeton for an eight-forty class, so I suggest that we take a five-minute break, and then that the few people who want to ask more questions do so afterward."

At that time there were perhaps two hundred people in the room; few had left. To my surprise, pleasure, and exhaustion, after the break at least half of them simply reseated themselves in front of the auditorium, and the questions went on for another hour, until I was finally rescued.

On the drive back to the Dean's house, where I was to spend the night, my host began to ask rather strange, elliptical questions.

"Do you feel personally threatened," he asked, "by all these rejection slips? Are you ego-involved with success or failure in getting people to discuss this new concept?"

I laughed and said, "No. My professional career is based on my high-energy research and my teaching, and I've had no difficulty in getting my regular work published. It's just

288

becoming frustrating that I think I'm really on to something worthwhile, that could benefit people quite profoundly, but no one seems willing to print my suggestions."

"It was important for me to ask that," continued the Dean, "because now I can tell you that I was one of the reviewers who rejected your article. After tonight, I feel that I should write to the editor and tell him that I've changed my mind."

Even now, three years later, I recall the pleasure and relief that came to me as the result of the warmth and intelligence of the audience at Hampshire. There have been many more lectures at colleges, with some very enthusiastic responses indeed, but Hampshire was the first, and I can never forget it.

Later that autumn I spoke at Princeton, at a colloquium of our Physics Department, and one student, seeing the colloquium sign which said "Space Colonization" spoke to me on the day of the lecture. "That's an academic joke, isn't it?" he asked, "I assume you're going to lecture on relativistic space-time."

Long after that lecture concluded, a few people remained to discuss the ideas. One of them, Professor Freeman Dyson from the Institute for Advanced Study, which is conveniently located only five minutes away from our university, stayed longest of all, and his kind interest led to a correspondence that has continued. Professor Dyson had written, years earlier, on the subject of advanced civilizations and their probable development of technology for habitation in space. In fact, he suggested, a really advanced civilization might build space habitations which would form a complete sphere, using the light of their Sun so completely that only infrared radiation would escape the star system;[4] the corresponding spectral shift might be a clue to the existence of such an advanced society. Professor Dyson called to my attention that day the early work of J. D. Bernal, and also guessed that there might be early writings of Konstantin Tsiolkowsky which could be relevant.

Before I left for California at the end of December 1972, I was emboldened to make one more assault on the academic publication establishment. This time I used my experience from years of responsibility as a group-leader in high-energy physics. I knew that to gain acceptance of a new experimental proposal it was often necessary to talk personally with the members of the ruling committee. This time I would go to someone I knew, and introduce my ideas in private discussion rather than simply in writing. An old friend from graduate school days, Dr. Harold Davis, was editor of the magazine *Physics Today,* the non-specialized publication of the American Institute of Physics. I went to New York and talked with Hal, over lunch, and left him the latest version of my often-rejected manuscript. Many months later, after review and consideration, Hal wrote that *Physics Today* would accept my article, if it were rewritten yet again to answer many more detailed questions. The rewrite occupied what time I could spare during the academic year 1973–74.

In 1973, when I was at Stanford full time working on an experiment on the new, large storage-ring called SPEAR, at the Stanford Linear Accelerator Center, I gave a number of lectures at colleges on the West Coast: Cal Tech, Stanford, and the campuses of the University of California at San Diego, Los Angeles, Berkeley, and Santa Cruz. In most cases the response was enthusiastic, and word of the new concept began to spread by individuals. There began to arrive letters which started: "I didn't know about your lecture until after it had happened, but I've been told about it by my friend and I want to know more . . ."

One letter, arriving late in 1973, was from a very young student at M.I.T., Eric Drexler. Said Eric:

"When I came to the university I set about finding who was working on space colonization; it was so obvious to me that logically someone should be. I tried to see if it is really true that you can reach anyone in the world with a sequence of not more than five phone calls or interviews, and Prof. Philip Morrison here has advised me to write to you." Thus began what has turned out to be a most pleasant friendship, and in early 1974 that friendship was cemented by a visit to Princeton by Eric and a friend of his from Columbia University, David Anderson. Now there were three people brave enough to gather in one place and talk about space colonization! In February of 1974, knowing that the *Physics Today* article would come out in a few months, we felt that we might dare even to hold a small conference on the topic.

Eric Drexler, David Anderson, and I, together with a Princeton graduate student, Eric Hannah, set our meeting for an early day in May, just after classes were to finish. There was little opportunity to set up an elaborate meeting, but I felt that as a matter of principle it should be possible to obtain at least a little money to support the conference. After all, we had spent more than a hundred billion dollars on the Vietnam War, and we were spending about as much every year in welfare programs and unemployment benefits. It seemed that space colonization was at least relevant to the issues of conflict, of human welfare and of employment.

I began with the established foundations, and soon found that none were interested in taking any real chances. Foundations generally have quite narrow charters, and their managements are usually reluctant to stray from the path of their previous grants. Some foundations, indeed, have reputations for the support of new directions in research, but I soon found that "new directions" meant "not really new."

After trying a number of the foundations which had been recommended as open-minded, and been turned down by all, I was led to a very small and special organization, the Point Foundation of San Francisco. Its office was a tiny two-room shack on the roof of the Glide Methodist Church, reached (on a rainy San Francisco day) over duckboards laid across the roof from the stairwell. Point, as I was to learn, owed its existence to a successful invention by Stewart Brand: *The Whole Earth Catalog*. Profits from the sale of the catalog supported Point, and the organization of the foundation was designed expressly to encourage innovation and prevent lockstep thinking. No officer or employee of Point, not even the part-time clerk who typed its correspondence, might hold office for more than three years. Each of its six trustees was given a modest block of money at the beginning of every year, and

291

from that time on was free to spend that money for any charitable purpose in which he individually believed; there were no committees, no reviews, no requirement of unanimity or even of agreement between the trustees.

In the shack, with the rain pouring down outside, I found a comfortable pair of small offices, lined with books. Richard Austin, secretary of the Point Foundation, was waiting for me. Richard, warm and friendly, quite without the airs of a "foundation executive," expressed his personal interest, and we were soon joined by Michael Phillips, a trustee, educated as a mathematician and now a man of many interests. Over a good lunch at the nearby San Francisco-Hilton Michael expressed his willingness to support the conference, and did so with a grant of six hundred dollars—tiny by the usual standards, but sizable for Point. Wisely, Michael suggested that the money be given as a formal grant to the university, so that, in his words, "The Establishment will be forced to recognize the existence of your work by filling out forms and generating a lot of red tape. Within many institutions, that is the only reality that is understood."

As the time for the conference drew near, Michael's action brought forth a response which meant little to me at the time, but which turned out to be important. It had never occurred to me to think of publicity for the meeting; it seemed brave enough to hold it at all. When the grant came through, though, the red tape predicted by Michael was generated, and part of it was a notification of the grant. That was sent, automatically, to the university's office of public information. There Florence Helitzer saw it and thought of the possibility of a news release. I reacted to the suggestion negatively at first; it seemed that we were getting rather out of our depth. Finally I gave permission for Florence to write the release, and she did so. Only by that sequence of events was there any news coverage of the First Conference on Space Colonization.

Our meetings began with a private half-day in which the two Erics, David, Freeman Dyson, Professor Gary Feinberg of Columbia, George Hazelrigg of the Princeton Engineering School, Gerald Sharp and Bob Wilson of NASA Headquarters, and Joe Allen, scientist-astronaut from

Crystal Palace habitat;
residential areas in sunlight.

NASA-Houston, sat down together to organize the talks for
the public day which would follow. In the previous
months I had worked out the details for the mass-driver,
and had considered its possible usefulness as a reaction
engine. Since late 1972 my lectures had included a discussion of another type of lunar launching-device, the "Rotary
Pellet Launcher," which also seemed to have possibilities
as a reaction motor.

Eric Hannah and Bob Wilson were in general agreement
on the lift costs for a shuttle-derived heavy-lift vehicle,
because NASA had kindly provided us with documents on
shuttle performance and costs. It appeared that payloads
could be brought to L5 for a cost of nine hundred and fifty
dollars per kilogram or less, and that if the initial habitat
could be held to modest dimensions, the entire construction program for the first habitat might carry a price tag not
very different from that of the Apollo Project.

May 10, the opening day of the conference, dawned dark and rainy, but some hundred to a hundred and fifty people braved the weather to show up for the start of the session. Walter Sullivan, science editor for *The New York Times*, was there, as were reporters from a number of local papers. By then I was too much involved in conference details to worry about whether we were about to fall on our faces.

The day's sessions went well, and questions were generally supportive and interested. Joe Allen had brought with him, in his "personal transportation" T-38 jet from Houston, a short film of the experiments done by one of the Skylab crews during their rest day. There we saw for the first time such ideal freshman-physics experiments as the formation, in zero-gravity, of a drop of water several inches in diameter, gently oscillating at a low frequency from spherical to football shape under the action of surface-tension forces. There too we saw the effect of internal friction on a rotating container of liquid.

Our private meeting the night before had concluded with a dinner put on for the speakers by my wife, and the second day of the conference concluded with a cocktail party at our house. We unwound with relief: the conference had seemed to go well, and we expected to return to quiet calculation and to reflection on the additional numerical details turned up by the conference speakers.

After a weekend of packing and organization for a summer of work in high-energy physics, on the Monday after the conference my wife and I left for California. We stopped to visit a great-aunt in Denver, and there began to realize what we were in for. The British Broadcasting Corporation tracked me down and wanted an interview; it seemed that Walter Sullivan had written an article about the conference, and that the editors of *The New York Times* had chosen to put it on the front page of that morning's edition. Other networks soon followed, newspaper and magazine reporters were not far behind, and a wave of public awareness and interest began to spread. Even now that interest shows no sign of slackening; indeed, it seems to grow with each passing month. As we try to assess the reasons for that interest, we must realize that it began well before the publication of the first scientific paper on space communities (that was not to come until the following September) and more than six months earlier than the first work on direct economic/energy benefits. Apparently, there was something fundamental about the space-community concept which made sense to many people, even without detailed arguments and plans. I have tried to understand, on the basis of the letters I have received and the conversations that follow lectures, what are the main reasons for that immediate positive response. These are guesses, and they are no more than that:

1. During the past few years people have felt a sense of increasing confinement, a sense of shrinking horizons and decreasing options. Suddenly, the humanization of space has appeared as a possibility, and many people feel an immediate sense of relief and freedom as a result. A sense, perhaps, that there could be a future of wide horizons, new freedoms, and excitement.

2. The space program up to the present time, valuable as it has been in many respects, has left many people with the feeling that they were being asked to pay for an "elitist" ego-trip, to be enjoyed personally only by a tiny segment of men, each capable of almost superhuman feats of physical endurance, dexterity, and technical competence. The vicarious pleasure of seeing the Apollo astronauts set foot on the lunar surface gave way, very quickly indeed, to the feeling: "All right for them, but what's in it for me? Do I want to spend my tax money so that some guy can hit a golf ball on the Moon?" In the humanization of space many people see the possibility of their own direct personal participation in an adventure more exciting even than the great explorations of the past. Popular interest and support may come at least in part from those desires for freedom and for participation, because they are gut-reactions which reinforce the logical arguments.

As the articles on space communities began to appear, and interviews began to be heard, letters began to flow in, initially from the English-speaking world and later from every continent. From the start, two characteristics about the mail were reassuring: first, that the letters of support outnumbered those of opposition by a ratio of about a hundred to one. Second, that the letters in any way irrational constituted no more than 1 percent of the total. The typical letter was thoughtful, lengthy, and represented a considerable input of effort and study on the part of the writer. For those reasons, the mail could not be answered in a careless or standardized fashion: a careful, thoughtful letter demanded an answer of the same quality.

For more than a year I struggled to keep up with the mail myself, but the rate increased continually, the quality remained high, and by mid-1975 the burden became too great. Since that time many of the letters and requests for information have been answered by a group of volunteers, each expert in a particular research area. Some letters, especially thoughtful and helpful, I must still deal with myself, and when I read them I regret not having the time to answer all the mail: clearly there is an enormous amount of information, a great influx of worthwhile ideas, arriving every day, and it is a loss that I must now receive much of it only secondhand.

Occasionally we send out a brief Newsletter, noting new publications and alerting those interested to recent and future events.[5]

The next milestone in placing the ideas of space colonization before the public was passed in late August and early September of 1974, with the appearance of a letter to *Nature*[6] and of the long-delayed article in *Physics Today*.[7] Hal Davis, editor of the latter magazine, used a painting of a space habitat for the cover of the September issue, and that painting, by Walter Zawoijski, was reproduced later in other publications.

In late May 1974 Mrs. Barbara Hubbard, of the Committee of the Future, a citizens' organization, became aware of our work and called to express her enthusiastic support. At that time the response was necessarily a practical one. It was clear that the subject of chemical processing of lunar materials needed a great deal of work. The M. I. T. student, Eric Drexler, was to be free during the summer, and was interested in doing all that he could to advance our understanding of those problems. I invited Mrs. Hubbard to support our work through Eric Drexler, and she promptly did so, donating $1,000 of money that the committee could ill afford. Eric did a fine job that summer, and there have been few occasions when so much was accomplished on a research budget so small. It is some measure of the rapid growth in each area of research on space manufacturing that by 1976 a NASA-supported study was devoting the full time of six people to the same subject of chemical processing; only a year later a NASA-funded summer study counted fourteen people, many of them senior professionals, working on the same topic. In 1977 the first long-term continuing research grant in this subject area was initiated.

The publication of the *Physics Today* article brought the concept of space colonization into the open for review by some 15,000 professional physicists, certainly as large and as critical a review board as one could wish for. Naturally, there were attempts by several to look for flaws in the arguments, to find numerical errors, or to point out possible absurdities in the assumptions made. During the autumn of 1974 it was necessary to devote a great deal of time to answering each of those critiques in detail. Some ran to twenty pages of closely reasoned argument and calculation, and had to be answered in kind.

Qualitatively, it seems that the criticisms all arose from an association, improper in my view, of numbers appropriate to the present day with technical problems that would arise only after many years. Several reviewers, for example, calculated the rate of transport that would be required if space colonization were ever to permit significant emigration from the earth. They concluded that the required transport rates would be absurdly high. Those numbers are not absurd for the years 2010-2050, when they might have to be faced, and it is irrelevant that they could not be achieved in 1980.

The other disagreement occurring most frequently arose from a failure on the part of the reviewers to allow for gradual growth, over many years, of industry and capability in space. The construction station for the first space habitat would have an aluminum plant with a capacity of only a few thousand tons per year, but if industry in space were to double in capacity every few years, as is certainly not impossible, by 2050 its output would be very large indeed.

Late in 1974 negotiations had begun with the Advanced Planning Division of the Office of Manned Spaceflight at NASA Headquarters in Washington. After much effort, those negotiations concluded with the initiation on January 1, 1975, of a small grant from that agency to Princeton, for support of our studies. Eric Hannah, who had just received his doctorate, provided valuable assistance for several months, and at the end of that year an old friend, Brian O'Leary, eminently qualified both scientifically and in governmental experience, joined in the effort at Princeton. Dr. O'Leary had been responsible for the first Hampshire College lecture on the humanization of space.

Within the organization chart of NASA, based on the Headquarters group and on the eight major NASA centers distributed across the United States, the Ames Research Laboratory in Mountain View, California is charged with the responsibility of looking into advanced systems and concepts. In September 1974 I gave a colloquium-lecture at Ames, and for the first time met the director of the laboratory, Dr. Hans Mark (later Undersecretary of the Air Force in the Carter Administration). Dr. Mark, a physicist who spent the early part of his career in nuclear physics of a military nature, has the reputation of working at least six days a week, of always arriving at work at 7:30 A.M., and of leaving the laboratory in the evening only long after everyone else but the night shift has gone. It was a pleasure to talk to him, and we soon arranged that we would "bootleg" a brief but intense research effort on space colonization by choosing that as the topic of the 1975 NASA Ames/Stanford University Summer Study. That study, one of an annual series supported by NASA and held in cooperation with the American Society of Engineering Education, was already funded, and the director of the laboratory was free to choose its topic each year.

As a consequence of the *Physics Today* article, there were many requests for lectures during the 1974-75

298

academic year. There were more than fifty invitations, and in order to give the new ideas a full review before a critical and knowledgeable audience I felt it incumbent upon me to accept many of them. At one point there were Physics Colloquia at Yale and Harvard back-to-back on the Friday and Monday that defined one weekend, and a special meeting with an M.I.T. group on the Saturday night.

In retrospect, such a review was necessary, however exhausting, but in the following academic year it became necessary to be more selective. The concept of the humanization of space has now passed beyond the point at which a departmental colloquium is the most appropriate forum for its discussion. The topic is, by its nature, interdisciplinary, so lectures which reach a wider audience are preferable: talks within university or college lecture-series, lectures before professional societies, the larger companies and laboratories, and interviews which have more than local distribution. Any group sincerely interested in hearing about the new possibilities should have its questions answered, and I turn over to qualified colleagues those invitations which I cannot accept myself.

In October 1974 several people active in space studies were invited to the Goddard Spaceflight Center near Washington, to provide information to NASA's "Outlook for Space" Committee. That group, chaired by Dr. Donald Hearth, was charged with the duty of preparing for NASA a list of possible tasks in space during the remainder of this century. Krafft Ehricke, Bruce Murray, George Field, and I gave our views, and there for the first time I heard Dr. Peter Glaser, of the Arthur D. Little Company in Boston, speak of his ideas on the generation of electric power from solar energy in space, and its transmission to Earth by microwave beam.

Until that time, although vaguely aware of Dr. Glaser's ideas, I had dismissed them as impractical, on the assumption that the microwave transmission problem must be insurmountable. To my surprise I learned that great progress had already been made on microwave transmission, and that the chief problems remaining related to the logistics of tons, dollars, and lift from Earth to geosynchronous orbit. In the days following the meeting at

Goddard I did some calculations, and soon found that the construction of satellite solar power stations out of lunar surface raw materials, in a high-orbital manufacturing facility at a space colony, would almost certainly solve the most serious problems then facing Dr. Glaser's concept. The calculations formed the basis of an article sent to the magazine *Science* in late December of 1974. Acceptance was prompt, and after extensive updating and revision the article was published on December 5, 1975.[8] The editors placed it first in the issue, and used as a cover painting a view of an early space habitat.

In early 1975 much of our effort was devoted to the organization of a second conference. The situation had changed greatly within a few months. The first conference had been informal, unofficial, and supported by a tiny grant from one small foundation. The second was an official Princeton University Conference, supported also by special grants from NASA and the National Science Foundation, and co-sponsored by the American Institute of Aeronautics and Astronautics (AIAA), the professional society of the aerospace field. The Conference was to last two and a half days and to consist of some thirty invited papers.

In the organization of the second conference, it was important to maintain a high level of professional expertise and seriousness. A year earlier, we had been a small, happy band of revolutionaries; now, with increasing recognition by professional and governmental bodies, it was both desirable and necessary to adopt a conservative and pragmatic approach. As a title I chose "Princeton University Conference on Space Manufacturing." Then, too, there was the small but important detail of the conference announcement itself. The essentials of the space-community concept did not depend on any particular choice of geometry for the initial habitat, and to emphasize that fact I chose for the cover of the announcement a photograph not of a habitat design but of the Woodrow Wilson building at Princeton. That building, soaring and modernistic, is graced by a reflecting pool and a fountain, and was to be the site for the conference itself.

The 1975 conference was counted a great success. Talks

on the economics of rocketry were supplemented by explorations of the legal, historical, psychological, and humanistic aspects of space habitation. The application of space manufacturing to the solution of the energy crisis on Earth received strong emphasis. On the last morning of the conference four summary talks were given, one for each of the preceding half-day sessions. Dr. Jerry Grey, an officer of the AIAA and formerly a professor of Aerospace Engineering at Princeton, gave the first summary talk. He was followed by Dr. John Billingham, chief of Life Sciences at the Ames Laboratory, and by Dr. Albert Hibbs, of the Jet Propulsion Laboratory at Cal Tech. JPL had conducted many of the spaceprobes to the Moon and planets.

As had happened a year earlier, there was strong response by the media to the conference, and a new round of articles and interviews began. Soon after the conference ended, I left with my wife for a brief ten days of complete change of pace: camping in a tent-trailer at a small grass airfield in Pennsylvania, where I could fly my sailplane and learn also to fly powered airplanes. As we look back on that brief period it seems an oasis of calm and peaceful pleasure in a year that had been far too intense at most other times.

The 1975 Ames Study was the first of several, but the only one funded through the ASEE program. On the first day of that study I turned over to the participants, to be copied for their use, all my notes and calculations from the nearly six years of research into space colonization. Soon afterward, copies of the papers from the 1974 and 1975 conferences were also sent out from Princeton.

According to the terms of the ASEE studies, the purpose of the summer was to be primarily educational; there was no requirement that the participants carry out a "most probable" or "most economical" design exercise, and no constraint that the study follow a preconceived plan. The participants chose the following problem: to design the elements necessary for the establishment of a colony in space, to house and maintain 10,000 people. Early in the study it was decided by the group that the primary goal would be not the return of energy or profits to the Earth, but the design and construction of a habitat able to house

301

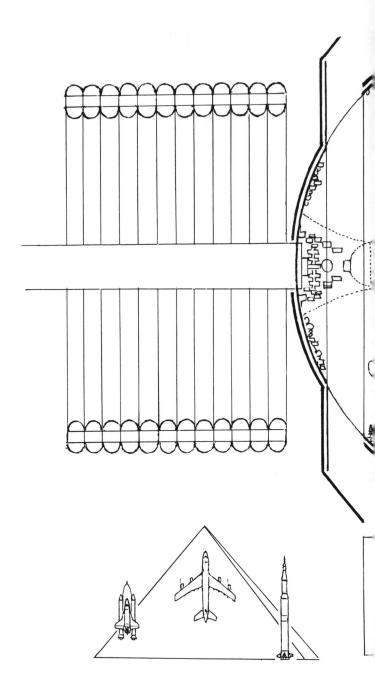

Detail of Bernal sphere with size comparisons.

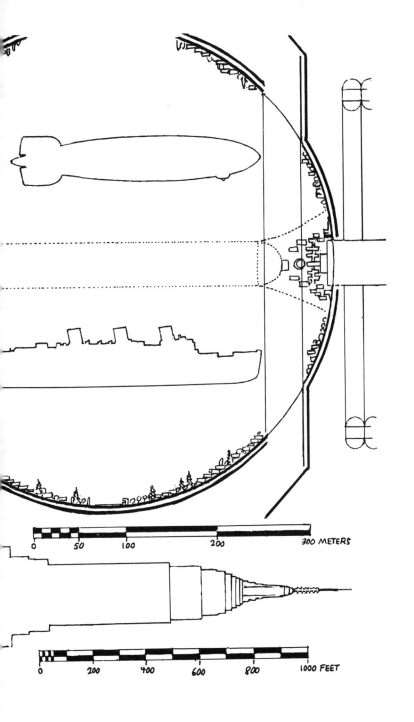

0   50   100   200   300 METERS

0   200   400   600   800   1000 FEET

Torus interior.

permanently a random sample of the population, including people with some degree of medical problems, pregnant women, children, and those unusually susceptible to motion sickness. That choice constrained the details of habitat design from that point on, but led to a thorough and valuable exploration of the wheel (torus) design. In any design study at an early phase of a new project, the most useful items are those of the greatest generality, least tied to particular design choices. From that viewpoint, particularly useful accomplishments of the 1975 Summer Study may turn out to be work done in the areas of productivity (in tons/person-year), of intensive agriculture, of closed-cycle ecology and of chemical processing.[9] The most significant single discovery was made by Dr. Eric Hannah, who tracked down the most informative articles and reports on the cosmic-ray intensity at a distance from Earth.

Subsequent studies were quite different in funding and goals: they were supported by NASA headquarters with the aim of developing near-term, practical routes to space manufacturing, and emphasized the production of satellite power stations and other useful high-orbital products. As such, they considered workforce habitats as necessary components of a total system, rather than as ends in themselves. Their conclusions as to habitat design centered on modular units and on the efficient use of structural and shielding mass. The spherical "Island One" design, quite sparing in its requirement for mass, survived as a likely candidate for a space settlement. The banded torus or "hatbox" design seemed most efficient among shielded habitats with a great deal of land area for their structural mass. Among geometries chosen for best use of volume rather than area, the sphere again proved best, according to a 1977 study of comparative habitat design. Implicit in these later studies was the assumption that a workforce of a few thousand people could be tested and selected, to reject those few unfortunate individuals who might be extraordinarily sensitive to inner-ear disturbances caused by rotation. The torus design, several times more massive than Island One for the same usable area, did not appear in these later studies to be competitive from a cost viewpoint with the more advanced designs.

In the middle of the 1975 Summer Study I was called to testify before Congressman Donald Fuqua's subcommittee of the U.S. House of Representatives' Committee on Space Science and Applications.[10] The reception was cordial, and a good block of time was allotted for discussion. Naturally, I emphasized the energy and economic benefits of high-orbital manufacturing, the Representatives present being far more interested in those aspects than in the longer-term philosophical issues of what Krafft Ehricke has aptly called the "Extraterrestrial Imperative." 1975 was a year of great progress. At its start, though public interest in space colonization was already considerable, there was still almost no work in progress on the topic except at Princeton. By the end of the year, active groups of students and faculty at universities such as M.I.T. and New York Polytechnic Institute had begun research on a volunteer basis, and each of the participants in the 1975 Summer Study had brought to his home university an interest and enthusiasm for further work, and for defending the basic ideas before lecture audiences. Citizens' groups independent of Princeton, notably the L5 Society,[11] had formed to supply information on the new possibilities and to publish commentary in newsletter form.

During early 1976 there were two developments of particular significance. First, notwithstanding a very tight NASA budget, a decision was made by that agency to fund a special study during the summer of 1976. This study took place at the NASA Ames Laboratory, and with the cooperation of the Ames directorate concentrated on three key technical subjects: the mass-driver, the chemical processing of lunar soil to obtain its oxygen, metals and clear glass, and the evolution, within the constraints of cosmic-ray shielding and an acceptable physiological environment for the construction work force, of a first construction station at L5 into Island One.

We were particularly fortunate in assembling for that study a very high-level team of aerospace professionals, each of whom arrived armed with documents and calculations from many years of practical experience in scientific and engineering problem-solving. Aided by an excellent group of students, within hours of their arrival they were hard at work on the three technical assignments, and in the second week of the study they were assisted by top-level specialists brought in as consultants.

The sensation that we were on the right track was increased by several developments early in the 1976 study. Each of the specialists expressed strongly the opinion that the critical numbers assumed for the work so far, and quoted in this book (mass-driver acceleration and efficiency, HLV lift costs, lunar power-plant mass, etc.) were too conservative and could be improved substantially without great technical risk. Professor Henry Kolm of M.I.T., leader of a group which had carried the magneplane concept through the level of successful tests on a superconducting-coil model comparable in size to a mass-driver bucket, brought detailed information on magnetic-levitation research now being carried out by programs each at a level of more than $100 million per year in Japan and Germany. On the basis of the assembled technical expertise, it appeared that the mass-driver could

306

operate at an efficiency of 80 to 90 percent, and could achieve accelerations of well over 100 g's, instead of the 29 g's assumed in my early calculations.

Dr. James Arnold of the Jet Propulsion Laboratory of Cal Tech, deeply involved in plans for a lunar polar orbiter spacecraft which may receive the significant title "Prospector," considered it highly probable that permanently shadowed areas on the Moon contain large deposits of hydrogen, carbon, and nitrogen in the form of ice and other compounds.

Dr. Brian O'Leary followed the scientific trail of a special class of asteroids known as the Apollo/Amors. Unlike the main-belt asteroids on which my economic calculations of Chapter 11 were based, the Apollo/Amors are separated from L5 by velocity intervals of as little as two or three kilometers per second, rather than the main belt's ten. By fortunate coincidence, only a few days after our 1976 study began the first known Apollo/Amor of carbonaceous type, rich in carbon, nitrogen, and hydrogen, was discovered.

As in 1975, I had to leave the study briefly to visit Washington, in this case for a meeting with Dr. James Fletcher, Administrator of NASA, and his deputy, Dr. Lovelace. At Dr. Fletcher's request I prepared, in cooperation with the study group, a list of more than a hundred research topics on which work is necessary if the concept of space manufacturing/industrialization is to be brought, in Goddard's words, from "the hope of today to the reality of tomorrow."

In a second 1976 development, the space-community concept was chosen by NASA as one of four major themes displayed in the form of exhibits at the Third Century America Exposition at the Kennedy Spaceflight Center in Florida for three months during the summer of 1976. This exhibit was moved to the California Academy of Sciences in San Francisco for six months in 1977, and subsequently has been shown at other large museums.

At the invitation of the Massachusetts Institute of Technology, I spent the 1976–77 academic year there as the Jerome Clarke Hunsaker Professor of Aerospace, while on sabbatical leave from Princeton University. It was a most productive year, and in October 1976 I was able to deliver to NASA the articles which summarized the 1976 Summer Study. Subsequently this set of articles was chosen as a volume in the series "Progress in Aeronautics and Astronautics" of the American Institute of that name, and after peer review was published in 1977.

Late in 1976 I became interested especially in the possibilities for development of mass-drivers, and in the spring term of 1977 gave a series of four seminars on mass-driver theory, the series being called "Spaceflight via Maxwell's Equations." During that entire year it was a special pleasure to work in close cooperation with Professor Henry Kolm of M. I. T. Together we designed the mass-driver model described in Chapter VIII, and the model was built during the first months of 1977 by Dr. Kolm and a group of student volunteers.

While at M. I. T. I considered the application of the mass-driver principle to a reaction-engine, capable of lifting accumulated shuttle-payloads totaling many hundreds of tons, from low Earth orbit to geosynchronous or lunar orbit. As the seminar-series continued in 1977 I found that such a reaction engine could perform much better than the best chemical rocket, would itself be light enough to be transported into orbit by only some four to six shuttle payloads, and could use as reaction mass the otherwise wasted material of the shuttle external tanks, which in NASA's original plan were to be discarded on each flight. Using the latest data from the 1976 study, in addition to the new insights being gained from the new theoretical developments, it appeared that the cost of reaching the "ignition point" in space manufacturing could be reduced substantially below the older $100 billion figure. That work, described in Chapter VIII, formed a starting point for the planning of time-schedules that was one of the activities of a 1977 study. It also suggested that the development both of closed ecological systems and of large monolithic habitats, previously seen as essential prerequisites for space manufacturing, could in fact be deferred to a later stage, well after the achievement of a high level of productivity.

In 1977 progress toward the humanization of space became still more rapid: the Princeton Conference of that year, held with co-sponsorship from AIAA, government agencies, and the General Electric Corporation, brought together nearly two hundred people. At the conference, the mass-driver model built at M. I. T. under the supervision of Dr. Kolm was demonstrated. Significantly, even that first model, built on a near-zero budget, showed an acceleration of more than thirty gravities, higher than I had once considered the ultimate for mass-drivers. By then, regular NASA support for a continuing program of research on that promising new concept had been obtained. At about the same time continuing research into the chemical processing of lunar materials was funded, and several NASA centers in addition to the Ames Laboratory began investigating the possibilities of their use.

With support from NASA centers and Headquarters, in 1977 another study was held, more than four times the size of the last. One of its tasks was the construction of a research plan with several options, aimed toward a program which could realize space manufacturing within the 1980s. The group concluded that within the launch-vehicle constraints of the space-shuttle era it should be possible to "ignite" space manufacturing through the "bootstrap" approach; in its scenario the first lift of equipment could begin as early as the mid-1980s, and substantial payback from manufacturing in space, at the level of many billions of dollars per year of income, could occur by the early 1990s. The 1977 study group concluded that an investment of roughly $60 billion, comparable to Apollo in 1977 dollars, would be enough to do the necessary research and development and pay the construction, salaries, and lift costs to the point where the program would become self-supporting. By the time the study ended the participants felt strongly that an augmented program of continuing research was urgent. Governmental decision delays appeared to be the most serious limits to the speed of progress; the technical results all looked better than earlier, though it was still possible that there might be hidden "show stoppers."

In 1977 also the Universities Space Research Association, with 55 member universities, completed the assembly of an advisory panel for a Task Group on Large Space

Structures; setting a precedent in USRA practice, this panel included representatives not only from engineering and science, but from the electric utility industry, the labor unions, and the investment community.

Directing these studies and chairing the USRA Task Group would certainly have become an impossible task for me, were it not for the fact that during those years a highly competent and dedicated group of friends and co-workers joined in the work. By now we are reaching that very productive kind of cooperation in which it is often impossible to identify any single individual as responsible for constructive new ideas.

Through the generosity of interested friends, in 1977 a supporting organization, the Institute of Space Research, Inc., was formed in Princeton. Able to accept gifts on a nonprofit tax-exempt basis, the Institute[12] aids our work in several ways, particularly by funding secretarial and other help to cope with the thousands of inquiries received each year in connection with the humanization of space. The officers of the Institute serve without compensation.

The demand for new informative articles both in this country and abroad continues, and in what little time can be spared from the continuing research there are lectures and interviews to be given. Dialog with governmental and corporate figures is frequent. We are not yet so far along that any single agency is ready to go out on a limb and grant the support that would be needed for intense full-scale research, but progress in acceptance and support of the new ideas even in a single year is so great that it would have been unthinkable a year before. Truly we may say that the humanization of space now appears as one of the most likely, as well as perhaps the most exciting and rewarding, of the possibilities open to humankind in the last quarter of the twentieth century.

In closing this account of the way in which the concept of the humanization of space began and has survived its earliest years, it is a particular pleasure to acknowledge the ideas and support of a number of friends. Although it is not possible to thank all who have helped in this work, my thanks go particularly to:

George Pimentel, John Tukey, Brian O'Leary, and Freeman Dyson, who encouraged this work from its
beginnings.

Janet, Roger, and Ellie O'Neill, who as children contributed their ideas and their encouragement.

Harold Davis, whose willingness to consider new possibilities with an open mind led to the first publication on
this topic.

Eric Drexler and Eric Hannah, whose interest and drive
were responsible in large part for the 1974 conference, to
Bob Wilson, Joe Allen, and Gerald Feinberg for their con-
tributions to it, and to Stewart Brand and Michael Phillips
who supported it.

The late Margaret Mead and John Stroud, who understood and worked toward many of these conclusions as
early as 1960.

The late Wernher von Braun, whose highest goal
through his working lifetime was the human movement
into space.

Krafft Ehricke, whose originality and drive can be seen
in ideas relating to almost every area of development in
space.

To officials within the U. S. Federal Executive who supported the Princeton work consistently through difficult
years of Government cutbacks: Hans Mark, John Yardley,
George Deutsch, Robert Freitag, Stanley Sadin, and Wayne
Hudson.

To individuals who as members of the Senate and House
of Representatives gave the High Frontier work their
encouragement: Senator Wendell Ford and Representatives Donald Fuqua, David Stockman, and the late Olin
"Tiger" Teague.

William B. O'Boyle and Barbara Hubbard, Lee Valen-
tine, Linda Ekman, and several anonymous donors, for
substantial gifts which made possible the establishment
and growth of the Space Studies Institute.

David Simpson and Erin Medlicott, for consistent sus-
tained work over a period of years in the buildup of the
Space Studies Institute.

The many thousands of Senior Associates and Sub-
scribers of the Space Studies Institute, who make possible

311

its substantial support of High Frontier research.

Stephen Cheston, James Arnold, Gerald Driggers, David Criswell, and Henry Kolm, whose professional work in science, engineering, and public affairs over a period of years in support of High Frontier development has been high in quality and unfaltering in commitment.

Robert Heilbroner, for permission to quote from "An Inquiry into the Human Prospect."

Isaac Asimov, whose articles and lectures eloquently support the human movement into the High Frontier.

And finally to my wife, Tasha, for making all problems lighter and all joys so much the greater.

APPENDIX 2

PERSPECTIVE—THE VIEW FROM 1988

In the twelve years since this book was written we have learned much about space development. What we have learned reinforces the conclusion that pioneers will migrate outward to the free space of our solar system. The first space settlers will leave Earth primarily to take up jobs in specialized new industries.

The largest new industry in space will meet our civilization's growing need for safe, economical energy. Serious accidents at nuclear power reactors in recent years, new data on the dangers to the biosphere of continued burning of fossil fuels, and the continuing lack of a practical near-term alternative Earthbound energy source confirm that the constant, reliable, virtually unlimited energy of sunlight in free space will be the ideal power source for an expanding civilization. Power satellites built in space will supply clean energy to the Earth, improving the health of Earth's biosphere.

New energy resources are needed both because Earth's population grows and because that population demands industrialization in order to reach an adequate living standard. The pressures of a growing population on diminishing planetary resources are even more apparent now than in the 1970's. The human population has increased in close accord with the predictions of Chapter 2, and Third-World forests are being stripped for fuel at a rate even more frightening than was foreseen in 1976. Warfare remains endemic on our crowded world.

Our knowledge of space resources, and of the engineering necessary for the High Frontier, have both advanced greatly in the past twelve years. Much of that research and development, which is outlined at the end of this chapter, was made possible by charitable donations to the Space Studies Institute. The material resources of nearby space, once thought to be confined to our Moon, have now been shown to include Earth-approaching asteroids undiscovered until the last decade. And each of the engineering developments necessary for the High Frontier has been demonstrated during these twelve years at least to the proof-of-concept level.

Progress in space development is uneven among the nations because it still depends mainly on political choices. Economic market forces are yet to have serious impact in the space arena. But when they do grow in strength, their effect will be enormous and will greatly accelerate progress, because private ventures driven by market forces tend to be much quicker than governmental programs. The reasons for that observation are fundamental. Governmental programs are paced by annual appropriations. A reduced appropriation or a budget overrun is accommodated by stretching out

314

the program. Within an agency there is little incentive for rapid completion of a program, but there are many incentives against innovation. Workers in a governmental agency therefore think in terms of decades and are relatively unconcerned about delays. Agencies tend to favor long, expensive programs which guarantee job security. In the civil space program the Apollo program was a unique exception. It was exempted from all normal bureaucratic rules and attracted a unique set of people, many of whom left NASA as soon as the Apollo mission was accomplished.

On the positive side, the same characteristics which make governmental agencies ill-suited to rapid completion of goal-oriented programs make them highly appropriate for sponsoring the long-term development of basic new technologies.

By contrast, in a private venture there is constant, intense pressure toward rapid, successful completion, because investors have put their money in the venture in the expectation of a high return. Until completion they cannot recover their investments and begin to earn profits. Delays reduce the value of profits when the profits are discounted to the year of investment. Workers in a private venture are therefore driven to complete programs and reach profitability quickly. For the same reasons, it is virtually impossible for private companies, except the very largest, to sponsor the development of basic new technologies.

My own experience in the private sector in recent years reinforces those conclusions, and confirms that of others who founded new enterprises. It can be summed up in three principles for goal-oriented ventures, public or private:

1) Keep it simple.

2) Avoid, if at all possible, developing new technologies or stretching old ones. Instead, assemble building blocks of existing technology in such a way as to build a new capability that serves a real need.

3) Keep your options open. Be ready to exploit new opportunities when they appear. Move fast.

The nations that followed those principles achieved success in the space race; those that did not fell back. In that race the most profound changes between 1976 and 1988 were the loss of position by the U.S., the steady progress of the U.S.S.R., and the development of strong, successful, independent space programs by China, Western Europe, Japan and India. It is to the credit of the U.S. that the last three of those programs benefited from significant transfers of NASA's technical expertise. The Soviet Union pursued a consistent long-term program aimed at the occupation of space and the settlement of the Inner Solar System. It established a series of records for the duration of manned spaceflight, and pioneered in growing food crops in space stations to feed its crews in orbit. It developed and flew a fully automated space plane larger than the U.S. shuttle, designed for re-entry and landing as a glider. To support its many launches the U.S.S.R. kept in production its expendable rockets, which built up a long record of reliability. In 1987 the first of the Soviet Union's new "Energia" rockets made a successful flight. The Energia's ability to lift mass to orbit (120,000 pounds in the first version of the Energia) dwarfs that of any lift vehicle since America's abandoned Saturn V series.

The space history of the United States in the same period was checkered. Spaceprobes were dispatched to Mars and to the outer planets. But after the great successes of the Apollo moon landings and of Skylab, the remaining unused Saturn V rockets were deliberately destroyed. As a result of a policy decision, production was halted for all other expendable rockets usable for American civil space launches. All of America's civil space activity, and much of its military program as well, were made to wait on the Space Shuttle, an experimental vehicle that stretched the arts of rocketry and of hypersonic aerodynamics. During the long drawn out development of the Shuttle, the habitation of near-orbital space, begun so promisingly with Skylab, was abandoned entirely. In 1973 the U.S. space program had been fifteen years ahead of all others. By 1988 that lead had been thrown away.

History teaches us that a strong, expanding frontier movement can endure only if it has a sound economic driver. The driver for the Westward movement of our 19th Century was the wealth of resources on successive new geographical frontiers. Beyond Earth, the nearest frontier within our solar system is low Earth orbit, and it is relatively barren. Solar energy is only available half-time there. Low orbit has no material resources, with one important exception, not yet realized. That is the external tanks of the Space Shuttle, which could be placed in orbit at almost no cost in payload.

Other than the external tanks, the nearest source of materials is the Moon, and the next nearest, typically 1,000 times as far away, is the Earth-approaching asteroids. Lunar and asteroidal materials have great value not because they are different from elements found on Earth, but because of their energy of height. They are at the top of Earth's gravitational mountain, while we are at the bottom. Because of that difference, it takes less than five percent as much energy to lift materials from the Moon into space as it does to lift equal amounts from the Earth.

Materials at the top of Earth's gravitational mountain will therefore go into products needed in space. When a large number of people are living in space, it will be clear which of the products they use will be made in space: all of them, except those which are so light, and so labor intensive in their manufacture, that Earth's industries can supply them competitively in spite of transport costs.

For the period before then, our knowledge in 1988 confirms the conclusion of Chapter 9: the main space industry will be constructing solar power satellites. While I wrote in 1976 that nuclear power would not be popular, the reality of 1988 goes far beyond my prediction. In the United States, virtually all construction of nuclear power plants has been halted by environmental and anti-nuclear protests, and there is considerable political pressure even to shut down nuclear plants that have been operating for many years.

While the concept of solar power satellites was invented (by Dr. Peter Glaser) in the United States nearly 20 years ago, the country that has taken the concept and run fastest with it is the Soviet Union. The USSR plans to have a pilot model solar power satellite in high orbit by the turn of the century, and to build a full-scale power satellite within another ten years.

Japanese scientists and engineers are also investigating solar power satellites intensively. That is logical, because Japan has no domestic sources of oil, and the Japanese are not enthusiastic about nuclear power. Observing Japan's track record in steelmaking, shipbuilding, automobiles and automated manufacturing, and also observing the rapid build-up of Japanese space capabilities during the past dozen years, it seems probable that Japan will be a formidable competitor in the construction of solar power satellites when that industry is established.

The size of the potential market confirms the estimate of Chapter 9. That estimate, made in 1976, was conservative in its assumptions about growth outside North America. Two realities of the past decade now reinforce the conclusions of Chapter 9: one is that "zero population growth" won't occur anytime soon. The world population has now passed five billion and is still growing. The second is that the "Confucian Economies" of the Western Pacific rim have grown rapidly, and are now among the world's great industrial powers. Therefore, the worldwide market for new and replacement electric power plants, a market which could be filled by solar power satellites, appears to be of the order of 400,000 megawatts per year, worth $0.4 trillion, almost 10% of the U.S. Gross National Product. That market is especially attractive because it can be predicted decades in advance.

Much of the Space Studies Institute's effort during the past decade has gone into detailed engineering and financing studies relating to the High Frontier. Those studies concentrated on achieving a self-sustaining, productive civilization in space at minimum cost and in minimum time. It became clear that the earliest productive facilities in orbit and on the Moon should be compact, and modular in form for easy replication. They should be built and tested on Earth, and then emplaced by unmanned rockets. Once in place, those facilities should be operated remotely by people at control consoles on the Earth, because astronaut/cosmonaut working time in space is very expensive.

Fortunately, the Moon is large and close to us, and also is locked by tidal forces to show the same face toward us all the time. Because of that good fortune, equipment on the near side of the Moon can be operated remotely ("teleoperated") from a single control station on Earth for more than eight hours at a time by a single direct line-of-sight radio link. Existing commercial communications satellites could link the Earth control station to three steerable antennas, 120 degrees in longitude apart. In that manner equipment on the Near-side of the Moon could be controlled continuously, 24 hours per day. The same simple geometry for communications could be used for any facility in high orbit.

The closeness of the Moon is vitally important for tele-operation, because of the fundamental limit of the velocity of light. A continuous television view of lunar operations can be received on Earth, and radio commands can be sent back to operate machinery on the Moon, with a round-trip time lag of only 2.7 seconds. The corresponding figure for any other planetary body is about 1,000 times as long: from 14 to 50 minutes for radio signals making the round trip to Mars, for example. Teleoperation is relatively easy with a 2.7 second delay, but nearly impossible when the delay is many minutes.

Vital concepts that have been quantified in the past decade are bootstrapping and the self-replication of machinery. Bootstrapping means living off the land, using materials found in space to reduce the quantity of materials needed from Earth. While that has been central to the High Frontier concept from the start, actual measurements have now yielded numbers that we may use in cost estimations. A solar-powered mass-driver on the Moon, operating only when the Sun is high in the lunar sky, can transport about 80 (eighty) times its total weight in lunar material each year to a precise point in high orbit.

Certain industries in space will depend on chemical processing plants to separate the lunar soils into pure elements (mainly oxygen, silicon, aluminum, iron and magnesium.) Measurements by Dr. Robert Waldron of Rockwell International on the chemical reactions necessary for that separation have shown that a chemical plant can process about 100 (one hundred) times its own weight in lunar materials each year.

In recent years "short-cut" processes have also been found, to extract particular elements or compounds of special value without going through the complexity of full chemical separation. Several of them are for separating oxygen, the "gasoline of space," which is the main component of rocket propellant. Another looks toward the magnetic separation of iron granules, which are plentiful in the lunar soil as the result of millions of years of bombardment of the Moon by meteors. Magnetic separation is a simple process. Once the iron is separated, it can be formed by pressure and heat, an industrial process called sintering. Sintered iron products are strong and can be made in precise shapes with closely controlled dimensions.

Still another process now undergoing practical testing is the forming of glass-glass composites. Thanks to work by the composites expert Brandt Goldsworthy, it has been found that the lunar glasses are separable by simple mechanical methods into two kinds, one melting at a low temperature, the other at a high. Goldsworthy has proven that one kind can be drawn into thin strong fibers, and that the other kind can be used as a comparatively soft matrix, similar in function to the organic resins used in fiberglass on Earth. His discovery opens the way to constructing habitats and other large structures out of materials readily available on the Moon, without the need for chemical separation of those materials.

Machines which can replicate most of their own components are of great advantage for establishing large scale industries in space. That concept was explored by John von Neumann and others many years ago, and it was realized in the early 1980's by such companies as Fanuc Limited, in Japan. That company manufactures industrial robots, in factories which employ those same robots operating under total control by computers, with no human attendance.

A small industrial seed, made up of a mass-driver on the Moon, processing plants on the Moon and in space, and general purpose fabricating shops ("job-shops") in both locations, can grow by self-replication into a mighty industrial power. Each of the three different modules should be small and designed for easy replication, and we should not attempt to replicate 100% of the parts. Large, simple, heavy parts needed in large numbers should be built in space. Small, complex parts like computers and precision tools should be made on Earth. The starting modules, each weighing around five tons, will be capable of manufacturing copies of themselves in about two months, even if only six percent of the lunar material they process ends up in their replicas. With a doubling time of two months, the industrial base will grow in steps of 1, 2, 4, 8, 16 ... Eight doublings will yield 256 times the initial capacity, and require only 16 months of time. Such a facility will be capable of processing 128,000 tons per year of lunar material. It can grow until its size is adequate to satisfy the demand for its products. If its main product is solar power satellites, and if that market is 400,000 megawatts per year of new and replacement power plants to supply energy for civilization on Earth, the facility can meet the need by growing another 2,000 times, requiring only eleven more doublings.

While the period 1973-1988 was a time of frustration, fundamental mistakes, tragedy and loss of purpose in the U.S. civilian governmental space program, there were also bright moments of achievement and clear vision in those dark years. One was the spectacular success of the Voyager scientific spacecraft, which photographed the outer planets and their moons, discovered previously unknown rings of matter around several of those planets, and continued in reliable operation toward the farthest reaches of the solar system. Another bright spot was the success, though long delayed, of the Space Shuttle in its first orbital flights.

Still another bright spot, in my opinion, was the appointment by the President in 1985 of a National Commission on Space, made up of 15 members and headed by Dr. Thomas O. Paine, a former Administrator of NASA. I had the honor to be on that Commission, serving with such interesting and likeable colleagues as the astronauts Neil Armstrong and Kathy Sullivan, the test pilot Chuck Yeager, and my old friends Dr. Luis Alvarez, Dr. George Field, Dr. David Webb and Dr. Paine.

With fifteen members, naturally the Commission could achieve agreement only on a conservative report. But it is remarkable what "conservative" meant in 1985-6 compared to ten years earlier. The very title of the Commission's report, "Pioneering the Space Frontier," connotes the expansion of the human race into a new habitat. Within the report, despite its conservatism, are found solar power satellites, space colonies like Island One, self-replicating factories in space, and mass-drivers. The fundamental concept of using materials found in space is highlighted by a Commission proposal for a "vigorous development of the technologies for robotic and teleoperated production of shielding, building materials, and other products from locally-available raw materials."

I am particularly pleased that our fifteen members followed an initiative of George Field and Frank White, to begin

321

the Commission's report with a nine-point "Rationale For Exploring and Settling the Solar System." The first of those points begins: "The Solar System is our extended home." Early in the report, a high priority is placed on developing a vehicle capable of carrying cargo and people to the Moon from low Earth orbit. The Commission's report advises a systematic, staged development of a transportation network from the Earth to the Moon and then beyond it. Soon after the report was published NASA carried out a companion study under the direction of the astronaut Sally Ride. I was pleased to see that the Ride study also recommended a systematic, staged development working outward, first to the Moon. Two years after those reports were completed the President of the United States endorsed their message in a statement on space, within which a human presence beyond Earth orbit was set as a fundamental goal.

Since the U.S. civil space program lost its drive and direction in the mid 1970's, there has been a great need for a responsible, continuing organization to carry forward the basic engineering research necessary for building space colonies and producing wealth from space. That organization, the Space Studies Institute, has been referred to earlier several times. It was founded in 1977 as a non-profit corporation, in essence a foundation. From its beginning it was supported by private citizens rather than by the Federal Government. After its first few years it also received modest but increasing support from private industry. A great many of the key developments since 1977 leading toward the High Frontier were funded by the Space Studies Institute; among them were:

- Computer-aided quantitative studies, published in aerospace journals, for reaching the High Frontier at low cost in minimum time, through bootstrapping and the partial self-replication of production machinery.
- Biennial Conferences on "Space Manufacturing and Space Colonies" co-sponsored by the Space Studies Institute and the American Institute of Aeronautics and Astronautics. Each of these conferences is recorded in a published Proceedings book, starting with a combined volume for 1974 and 1975, and continuing for Conferences held in all odd years since then. These proceedings contain papers in all the research areas needed for

the High Frontier, from engineering to psychology and architecture.

- Research on the design and cost of satellite power stations constructed primarily of lunar material. The reports based on that research show that more than 99% of the mass of a solar power satellite can be built of materials abundant on the Moon.

- A thesis by Scott Dunbar, carried out in the Physics Department of Princeton University, which indicates that asteroidal or cometary material may be trapped in the Earth's orbit around the Sun. It is very difficult to search for that material by way of telescopes on the Earth, because of the angle at which sunlight strikes it. The problem is much like that of searching for an airplane whose position in the sky is close to that of the Sun. The ideal device to find such material would be a small telescopic probe flown in the plane of the Earth's orbit around the Sun, on an orbit slightly inside the Earth's, so the Sun would be at its back.

- A program of measurements resulting in a detailed design for a chemical processing plant to separate lunar soil into metals, oxygen, silicon and other useful elements. That program was carried out by Rockwell International under the direction of Dr. Robert Waldron, through a contract from SSI. It was of special interest because none of the governmental studies which preceded it had checked with reality by measuring the actual chemical reactions needed. The SSI-sponsored study began by obtaining hard data through measurements, and developed detailed designs and performance numbers for the separation plants based on that data.

- The design, construction and testing of Mass Driver III, bringing the mass-driver concept to the level of certainty necessary for its inclusion in plans for manufacturing in space from lunar materials. Mass Driver I was built and tested in 1977. It showed an acceleration of 35 gravities. Mass Driver II was built in the Physics Department of Princeton University, during the years 1978-80, funded by grants from NASA and SSI. By 1981 the insights gained from seven years of calculation and testing led me to the concept of Mass Driver III. In the following year I wrote and checked a Computer Aided Design program for the machine, and concluded that mass-drivers could operate at 1,800 gravities, accelerating payloads from zero to a speed great enough to escape the Moon's gravity in hardly more than a tenth of a second. Dr. Les Snively, working on an SSI grant to Princeton, and aided by volunteers associated with SSI, built a working model of the first half-meter of Mass Driver III in time for its operation at the 1983 SSI/AIAA Biennial Conference. The machine worked just as expected, within one percent of the performance numbers predicted by the computer program.

SSI has sponsored a number of other research programs; those above are some of the longer ones, but others of shorter duration have also yielded engineering data vital to the practical realization of the High Frontier.

New knowledge gained through research on rotating environments, and data from record-breaking Soviet endurance flights in Earth orbit, now give us a much wider choice of scale for the Island One space colony that was envisioned in the 1970's. The Soviet data proves that a colony in space must rotate to provide the equivalent of gravity in order for its inhabitants to maintain perfect health. Other experimental data make it quite likely, though not yet certain, that the gravity of the colony need be only about half as strong as that of Earth, and that colony rotation rates of three to four per minute will be acceptable to most people. Within that range of parameters there are new and interesting possibilities. A colony providing half of Earth's gravity at its equator and rotating three times per minute would be 100 meters in diameter, would have the right land area for 500 people, and would require a pressure-hull mass of only 700 tons. The smallest scale colony of the spacious, visually open spherical design would be 60 meters in diameter, would house 150 people comfortably, and would require only 130 tons in its spherical pressure shell. The necessary metal for it could be obtained from external fuel tanks left in orbit from just four Space Shuttle flights.

I have noted that much of the research lifting us toward the High Frontier of space is being carried out by the Space Studies Institute. The Institute's "fuel" is many small private donations, not government grants. To provide more fuel, so that the Institute could fly faster and higher, I made an unusual arrangement in the course of establishing a private company. In 1982 a patent was issued to me for a new kind of satellite service, to provide accurate navigational positioning and two-way digital message transfer via satellites, for small terminals carried by hand or in vehicles. Shortly afterward I founded a company, the Geostar Corporation, based on the patent. The first operating orbital element of the Geostar satellite system for North America was launched on an Ariane rocket in March 1988, and the company began revenue service soon afterward.

My unusual arrangement was to allocate about 85% of Geostar's founding stock to SSI. The Institute paid $4,600 for that stock, and by 1988 the stock, though not publicly traded, was valued at more than $20 million. All of us who have helped to build SSI over its first decade hope that its Geostar stock will continue to gain value, giving the Institute the economic muscle it needs to make the High Frontier program a reality.

This chapter has been devoted to the daily work, much of it by the Space Studies Institute, that is enabling our climb to the High Frontier. We need, I believe, to lift our eyes above those daily tasks occasionally, to remind ourselves of the shared vision for which our work is done. Ultimately that vision will expand our physical, political and mental boundaries, from the confines of a single planet to the much broader limits of a race freely expanding its habitat throughout our solar system, and from there to the stars. Even the beginning of realization of that vision will bring profound benefits to our planet and its life:

> The sure survival of all the races of humanity, and of the plant and animal life forms we cherish as part of our Earthly heritage, in colonies dispersed throughout our solar system and beyond it.

> The preservation of the Earth and its fragile biosphere, as a place of great beauty, deserving our care and our nurturing, as it has nurtured us through our evolution.

> Opening a hopeful future for individual human

beings, with increasing personal and political freedoms, a wider range of choices, and greater opportunities to develop individual potentials.

Reducing the incidence of wars and the constant threat of wars, by opening a new frontier with virtually unlimited new lands and new wealth.

These are the worthiest of goals, and many of us have tried in our own ways to work toward them. We may take courage in the fact that by opening the High Frontier we will transform all four of those goals into reality.

ACKNOWLEDGMENTS

It is with deep appreciation that I thank the people who have made this edition of *The High Frontier* possible: Bettie Greber for coordinating the project, and for her tireless detective work in locating artists and finding original plates. Peter Thorpe for his excellent work in developing the cover concept and designing the jacket layout. Pat Rawlings of Eagle Aerospace for creating in the cover painting an image of life in an Island One habitat so exciting that we want to go there now. Rick Norman Tumlinson for proposing this new edition and for his continuing hard work toward making the dream of the High Frontier a reality. Bob Werb for dedicating a year of his life to learning the publishing business, and dedicating his own resources to accomplishing this project.

REFERENCES

CHAPTER 1

1. Lucian of Samosata, *A True History* and *Icaromenippus*, circa A.D. 160, English translations (respectively), New York: Murray, Scribner & Welford, 1880, and Oxford: Clarendon Press, 1905.
2. E. E. Hale, "The Brick Moon," *Atlantic Monthly*, vol. XXIV, October, November and December, 1869.

3. J. Verne, *Off on a Comet*, Paris, 1878.

4. K. K. Lasswitz, *On Two Planets*, Leipzig, 1897.

5. K. E. Tsiolkowsky, *Dreams of Earth and Heaven*, Moscow, 1895.

6. K. E. Tsiolkowsky, *The Rocket into Cosmic Space*, Moscow, Naootchnoye Obozreniye, 1903.

7. R. H. Goddard, "The Ultimate Migration," manuscript dated Jan. 14, 1918, The Goddard Biblio Log, Friends of the Goddard Library, Nov. 11, 1972.

8. R. H. Goddard, "2. Importance of Production of Hydrogen and Oxygen on the Moon and Planets," manuscript notes, March 1920.

9. H. Oberth, *The Rocket into Interplanetary Space*, Munich, 1923.

10. G. von Pirquet, articles, *Die Rakete*, vol. II, 1928.

11. H. Noordung (Potocnik), *The Problems of Space Flight*, Berlin: Schmidt and Co., 1929.

12. J. D. Bernal, *The World, the Flesh and the Devil*, London: Methuen & Co., Ltd., 1929.

13. O. Stapledon, *Starmaker*, London: K. Paul, Trench, Trubner & Co., 1929.

14. H. T. Rich, "The Flying City," *Astounding Stories*, August 1930.

15. F. Zwicky, "Morphological Astronomy," The Halley Lecture for 1948, delivered at Oxford, May 2, 1948, *The Observatory*, vol. 68, August 1948, pp. 142–3.

16. H. E. Ross, "Orbital Bases," *J. British Interplanetary Society*, vol. 8, no. 1, 1949.

17. A. C. Clarke, "Electromagnetic Launching as a Major Contributor to Space Flight," *J. British Interplanetary Society*, vol. 9, 1950, p. 261.

18. W. von Braun, "Crossing the Last Frontier," *Collier's*, March 22, 1952.

19. L. R. Shepherd, "Interstellar Flight," *J. British Interplanetary Society*, July 1952.

20. A. C. Clarke, *Islands in the Sky*, Philadelphia: John C. Winston, 1952.

21. I. M. Levitt and D. M. Cole, "Exploring the Secrets of Space," Englewood Cliffs, N. J.: Prentice-Hall, Inc., 1963, pp. 277, 278.

22. F. J. Dyson, "Search for Artificial Stellar Sources of

Infrared Radiation," *Science,* vol. 131, June 1, 1960, p. 1667.

23. V. P. Petrov, *Artificial Satellites of the Earth,* translated by B. S. Sharma and R. S. Varma, Ministry of Defense, Gov't of India, New Delhi: Hindustan Publishing, 1960, p. 217.

24. K. P. Osminin, "Questions of Economics and International Cooperation in Space Operations," XXVth International Astronautical Congress, Amsterdam, The Netherlands, Sept. 30—Oct. 5, 1974.

25. K. A. Ehricke, "Space Stations—Tools of New Growth in an Open World," XXVth International Astronautical Congress, Amsterdam, The Netherlands, Sept. 30—Oct. 5, 1974.

26. A. Berry, *The Next 10,000 Years,* New York: Saturday Review Press/E. P. Dutton & Co., 1974.

27. G. Harry Stine, *The Third Industrial Revolution,* New York: G. P. Putnam's Sons, 1975.

(For references 1-25, cf. R. Salkeld, "Space Colonization Now," *Aeronautics and Astronautics,* September 1975, p. 30, as reviewed by F. C. Durant III.)

CHAPTER 2

1. Carleton S. Coon, *The Story of Man,* New York: Alfred A. Knopf, 1954.

2. Sebastian von Hoerner, "Population Explosion and Interstellar Expansion," in *Einheit und Vielheit,* Göttingen: Van den Houck & Ruprecht, 1973.

3. J. C. Fisher, *Energy Crises in Perspective,* New York: John Wiley & Sons, 1973.

4. E. F. Schumacher, *An Economics of Permanence,* Institute for the Study of Non-Violence, Box 1001, Palo Alto, California 94302.

5. Population Studies #53, U.N. Department of Economic and Social Affairs, United Nations, New York, 1973.

6. Von Hoerner, *op. cit.*

7. *Ibid.*

8. P. A. Taylor, "World Population Conference 1974"; interview with Ansley J. Coale: *Princeton Alumni Weekly,* Oct. 22, 1974, p. 8.

9. David R. Safrany, "Nitrogen Fixation," *Scientific American*, October 1974, vol. 231, #4, pp. 64–81.
10. Fisher, *op. cit.*
11. Associated Universities, Inc. AET-8, April 1972.
12. Fisher, *op. cit.*
13. Jean-Jacques Faust, *L'Expresse*, Nov. 18–24, 1974.
14. Safrany, *op. cit.*
15. Associated Universities, Inc., *op. cit.*
16. J. McPhee, "The Curve of Binding Energy," *New Yorker*, Dec. 17, 1973.
17. Von Hoerner, *op. cit.*
18. Schumacher, *op. cit.*
19. Robert Heilbroner, *An Inquiry Into the Human Prospect*, New York: W. W. Norton, 1974. Page references: 134, 17, 88, 44, 43, 93, 110, 108, 26, 141, 140, 27, 136.
20. J. W. Forrester, *World Dynamics*, Cambridge, Mass.: Wright-Allen Press, 1971.

CHAPTER 3

1. *National Geographic*, January 1975.
2. Gerald Feinberg, *The Prometheus Project*, Garden City, New York: Doubleday, 1969.
3. David Hafemeister, "Science and Society Test for Scientists: The Energy Crisis," *American Journal of Physics*, August 1974.

CHAPTER 4

1. David R. Safrany, *op. cit.*
2. J. C. Fisher, *op. cit.*
3. G. Harry Stine, *op. cit.*
4. Lewis Beman, "Betting $20 Billion in the Tanker Game," *Fortune*, August 1974.
5. T. B. McCord and M. J. Gaffey, "Asteroids: Surface Compositions from Reflection Spectroscopy," *Science*, October 1974.
6. E. K. Gibson, C. B. Moore, and C. F. Lewis, "Total Nitrogen and Carbon Abundances in Carbonaceous Chondrites," *Geochimica et Cosmochimica Acta*, 35:599, June 1971, #6, pp. 599–604.
7. K. Tsiolkowsky, *Beyond the Planet Earth*, trans. by Kenneth Syers, New York: Pergamon Press, 1960.

8. K. Tsiolkowsky, "Selected Works," Moscow: Mir Publishers, 1968.

CHAPTER 5

1. G. K. O'Neill, "Colonization at Lagrangia," *Nature*, August 23, 1974.
2. G. K. O'Neill, "The Colonization of Space," *Physics Today*, September 1974.
3. "Multiple Cropping—Hope for Hungry Asia," *Reader's Digest*, October 1972, p. 217.
4. Richard Bradfield, private communication.
5. F. M. Lappe, *Diet for a Small Planet*, New York: Ballantine Books, 1971.
6. "Multiple Cropping . . . ," *op. cit.*

CHAPTER 6

1. G. K. O'Neill, "The Colonization . . . ," *loc. cit.*
2. Henry H. Kolm and Richard D. Thornton, "Electromagnetic Flight," *Scientific American*, October 1973.
3. Arthur C. Clarke, "Report on Planet Three," Signet #5409, New York: New American Library, 1972.
4. Richard Bach, Dedication to *Jonathan Livingston Seagull*, New York: Macmillan Co., 1970.

CHAPTER 7

1. "Meteoroid Environment Model—1969 (Near Earth to Lunar Surface)," NASA SP-8013, 1969.
2. G. Latham, J. Dorman, et al., "Moonquakes, Meteorites and the State of the Lunar Interior," and "Lunar Seismology," in *Abstracts of the Fourth Lunar Science Conference, 1973*; Lunar Science Institute, 3303 NASA Road 1, Houston, Texas 77058.
3. R. E. McCrosky, "Distributions of Large Meteoric Bodies," Smithsonian Astrophysical Observatory Special Report #280, 1968.
4. Morgan and Turner, eds., *Natural Environment Radiation Exposure*, New York: John Wiley & Sons, 1967.
5. G. M. Comstock, R. L. Fleischer, et al., "Cosmic-Ray Tracks in Plastics: The Apollo Helmet Dosimetry Experiment," *Science*, April 9, 1971.

6. *Wissenschaften 60,* 233, 1973.
7. John McPhee, *The Curve of Binding Energy,* New York: Farrar, Straus & Giroux, 1974.

CHAPTER 8

1. Edison Electrical Institute, *Statistical Yearbook of the Electric Utility Industry for 1973,* New York: Edison Electrical Institute, 1973.
2. G. D. Friedlander, Institute of Electrical and Electronic Engineers *Spectrum 12,* May 1975.
3. W. R. Cherry, *Aeronautics and Astronautics,* August 1973.
4. Report of the 1975 NASA-Ames/Stanford University Summer Study on Space Colonization.
5. H. Davis, *Proceedings,* 1975 Princeton University Conference on Space Manufacturing Facilities, Paper 1–6, New York, American Institute of Aeronautics and Astronautics (AIAA).
6. A. O. Tischler, *ibid.,* Paper 1–5.
7. G. K. O'Neill, "The Colonization . . . ," *loc. cit.*
8. A. C. Clarke, *Journal of the British Interplanetary Society,* vol. 9, 1950.
9. H. H. Kolm and R. D. Thornton, "Electromagnetic Flight," *loc cit.*
10. Kevin Fine, Eric Drexler, Bill Snow, Jonah Garbus (M.I.T.) Jon Newman (Amherst).
11. B. Mason and W. G. Melson, *The Lunar Rocks,* New York: Wiley-Interscience, 1970.
12. G. K. O'Neill, "The Low (Profile) Road to Space Manufacturing," *Astronautics and Aeronautics,* September 1977.

CHAPTER 9

1. D. Hayes, *Science,* June 27, 1975.
2. W. R. Cherry, "Harnessing Solar Energy: The Potential," *Aeronautics and Astronautics,* August 1973.
3. P. E. Glaser, Space Shuttle Payloads, Hearing, Committee on Aeronautical and Space Sciences, U. S. Senate, Oct. 31, 1973, Part 2.
4. W. C. Brown, Proc. IEEE, January 1974.
5. News release, Office of Public Information, Jet Propul-

sion Laboratory, California Institute of Technology, Pasadena, May 1, 1975.

6. G. L. Woodcock and D. L. Gregory, American Institute of Aeronautics and Astronautics paper 75-640, presented at the American Institute of Aeronautics and Astronautics/American Astronomical Society Conference on Solar Energy for Earth, April 24, 1975.

7. K. Bammert and G. Deuster, paper presented at the American Society of Mechanical Engineers Gas Turbine Conference, Zurich, April 1974.

8. R. E. Austin and R. Brantley, presentation at NASA Headquarters, Washington, D.C., April 17, 1975 (unpublished).

9. R. E. Austin, NASA-Marshall Spaceflight Center, private communication.

10. G. K. O'Neill, "Space Colonies and Energy Supply to the Earth," *Science,* Dec. 5, 1975.

11. G. K. O'Neill, Testimony before the Subcommittee on Space Science and Applications of the Committee on Science and Technology, U. S. House of Representatives, July 23, 1975. Superintendent of Documents, U. S. Government Printing Office, Washington, D. C. 20402.

12. G. K. O'Neill, Power Satellite Construction from Lunar Surface Materials, Testimony before the Subcommittee on Aerospace Technology and National Needs of the Committee on Aeronautical and Space Sciences, U. S. Senate, January 19, 1976; U. S. Government Printing Office, Washington, D. C. 20402.

13. Exxon Corporation, advertisement in *Smithsonian* magazine, April 1975.

14. G. K. O'Neill, "A High Resolution Orbiting Telescope," *Science,* May 1968.

15. R. N. Bracewell, "The Galactic Club," Stanford Alumni Association, Stanford, California, 1974, and *Nature,* vol. 186, 1960.

16. G. Cocconi and P. Morrison, "Searching for Interstellar Communications," *Nature,* vol. 184, 1959.

17. F. D. Drake, "Project Ozma," *Physics Today,* vol. 14, p. 140, 1961.

18. Ian Ridpath, *Worlds Beyond,* New York: Harper & Row, 1975.

19. R. N. Bracewell, *op. cit.*
20. Bernard M. Oliver, ed, "Project Cyclops," NASA Report #CR114445, 1973.
21. Carl Sagan, *The Cosmic Connection*, New York: Doubleday-Anchor Press, 1973.

CHAPTER 10

1. T. Taylor, "Propulsion of Space Vehicles," in R. Marshak's *Perspectives in Modern Physics*, New York: Wiley-Interscience, 1966.
2. R. Bradfield, "Multiple Cropping . . . ," *loc. cit.*
3. F. M. Lappe, *loc. cit.*

CHAPTER 11

1. H. S. F. Cooper, Jr., *A House in Space*, New York: Holt, Rinehart & Winston, 1976.
2. T. B. McCord and M. J. Gaffey, *loc. cit.*
3. C. R. Chapman, D. Morrison, and B. Zellner, "Surface Properties of Asteroids: A Synthesis of Polarimetry, Radiometry and Spectrophotometry," *Icarus*, vol. 25, 1975.
4. *Ibid.*
5. For turbogenerator combinations in the 1300 MW size range, the purchaser (TVA) in 1975 paid the builder (Brown-Boveri Corp.) $56/KW; (*Wall Street Journal*, Jan. 23, 1975). The corresponding cost per 500 KW is $28,000. In the text, the figure is doubled to include nonturbogenerator components of a complete power plant.
6. R. Heilbroner, *loc. cit.*, p. 140.

CHAPTER 12

1. Klaus P. Heiss, "Our R+D Economics of the Space Shuttle," *Aeronautics and Astronautics*, October 1971.
2. G. K. O'Neill, *Nature*, *loc. cit.*
3. Columbus lived from 1451 to 1506; Francis Drake, from 1545 to 1595; Michelangelo spanned the years 1475–1564, and Shakespeare 1564–1616.

APPENDIX 1

1. Charles Dickens, *A Christmas Carol.*
2. C. W. Allen, *Astrophysical Quantities,* London: Athlone Press, 3rd ed., 1973.
3. I have taken care *not* to check the details of this story with John himself; perhaps it was embroidered a little by the time it reached me, but I don't want to spoil it by giving qualifications.
4. Such an arrangement, referred to in an earlier chapter, has been called a "Dyson Sphere."
5. To be added to the mailing list for Newsletters, address: Professor Gerard K. O'Neill, Physics Department, Princeton University, Box 708, Princeton, New Jersey 08540. Newsletter readership has now grown to the point where it can only be handled on a subscription basis. See ref. 12 for address for inquiries on newsletter subscription rate.
6. G. K. O'Neill, *Nature, loc. cit.*
7. G. K. O'Neill, "The Colonization of Space," *loc. cit.*
8. G. K. O'Neill, Science, vol. 190, no. 4218, Dec. 5, 1975.
9. Space Settlements; a design study. NASA SP-413.
10. Space Colonization and Energy Supply to the Earth, Testimony of G. K. O'Neill before the Subcommittee on Space Science and Applications of the Committee on Science and Technology, United States House of Representatives, July 23, 1975, *loc. cit.*
11. L5 Society, 1620 North Park Avenue, Tucson, Arizona 85719.
12. Space Studies Institute, Box 82, Princeton, New Jersey 08540.
13. Partial listing of publications:
    Nature, August 23, 1974.
    Physics Today, September 1974.
    "Space Colonization and Energy Needs on Earth," *Science,* December 5, 1975.
    "Settlers in Space," *Science Year 1976,* September 1975.
    "Present Status of Space Manufacturing Studies," *Aerospace Magazine,* November 1975.

New York Times, May 13, 1974.

New York Times, June 12, 1975.

New York Times Magazine, Jan. 18, 1976.

Space Manufacturing Facilities/Space Colonies; Proceedings of the 1974 and 1975 Princeton Conferences; Amer. Inst. of Aeronautics and Astronautics, New York (published as a single combined volume).

"Engineering a Space Manufacturing Center," *Astronautics and Aeronautics*, October 1976, p. 20.

"The Low (Profile) Road to Space Manufacturing," *Astronautics and Aeronautics*, September 1977.

Space Manufacturing from Non-Terrestrial Materials: the 1976 NASA Ames Study (in press, as volume in series: *Progress in Aeronautics and Astronautics*, AIAA, New York, N.Y.).

Proceedings, Third Princeton/AIAA Conference on Space Manufacturing, May 9–12, 1977; AIAA, New York, N.Y.

INDEX

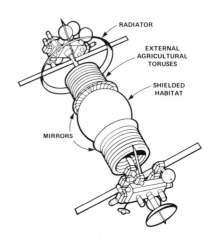

RADIATOR

EXTERNAL
AGRICULTURAL
TORUSES

SHIELDED
HABITAT

MIRRORS

YES, I want to help make the dream of space colonies and space manufacturing
ality. Count on me as a Sustaining Member of the Space Studies Institute.
further the Institute's critical research work here is my tax deductible gift of:

$200 □ $100 □ $75 □ $50 □ $25 □ $15 Student □ Other _____

ke your check payable to "Space Studies Institute."

Master Charge   □ Visa   Card No. _____

p. Date _____

nature _____

me _____

dress _____

y _____ State _____ Zip _____

donations are tax deductible.

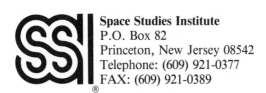

Space Studies Institute
P.O. Box 82
Princeton, New Jersey 08542
Telephone: (609) 921-0377
FAX: (609) 921-0389